NATURE'S ARCHITECT

The beaver's return to
our wild landscapes

JIM CRUMLEY

Saraband

Published by Saraband
Suite 202, 98 Woodlands Road
Glasgow, G3 6HB, Scotland
www.saraband.net

ISBN: 9781910192061
ebook: 9781910192078

Printed and bound by CPI Group (UK) Ltd, Croydon, CR0 4YY
on sustainably sourced paper.
Editor: Craig Hillsley
Cover illustration: Joanna Lisowiec

1 2 3 4 5 6 7 8 9 10

CONTENTS

To the beaver pioneers
JOHN LISTER-KAYE
and PAUL RAMSAY

and to DON MACCASKILL
for daring to dream

If no exaggerations had ever appeared in connection with the beaver, except those referring to its performances in felling trees, the stock of these alone would have been sufficient to damage the reputation of Natural History writers.

– Horace T. Martin, *Castorologia*
(W. Drysdale, Montreal, 1892)

Chapter 1

LIQUID ARCHITECTURE

The mother art is architecture.
  – Frank Lloyd Wright

Study nature, love nature, stay close to nature. It will never fail you.
  – Frank Lloyd Wright

The beaver was the original inventor of reinforced concrete. He has used it for a million years, in the form of mud mixed with sticks and stones...
  – Ernest Thompson Seton

Liquid architecture. It's like jazz – you improvise, you work together, you play off each other, you make something...
  – Frank Gehry

IT OCCURS TO ME from time to time that there are three things I might have become: a nature writer, a jazz guitarist, or an architect. Of these three, nature writing has been my day job (and often my night job too) for nearer thirty years than twenty now; my jazz guitar playing is something

1

of a well-kept secret, a handful of gigs a year where the first criterion to satisfy is that the audience outnumbers the band; and I have more chance of walking on water or flying unaided somewhere over the rainbow than becoming an architect because (a) I can't draw, and (b) having to think in three dimensions is usually at least one too many for me and sometimes two. Then I discovered beavers, for which architecture, inspired improvisation, and reworking nature into dazzling essays, are simply the stuff of life itself. In other words, all three of my preferred lifestyle alternatives co-exist within any one member of the beaver tribe in more or less equal parts. Working with nature can be a humbling experience for a mere mortal.

I love that phrase of Frank Gehry's − liquid architecture. It might have been invented for a book about beavers. Beavers build landscapes from scratch and from water that flows and timber that grows in the wrong place. All life is improvised liquid architecture, to which end water must be taught not to flow and timber not to grow, at least not there, but rather where the one can be purposefully redirected and the other artfully felled, reposed and metamorphosed into dam, lodge, bridge, larder, fridge; and water must not flow there but rather use this new canal, these new plunge pools, those channels one-beaver-wide for fast returns to deep water from the woods whenever danger manifests itself.

Frank Gehry is a bit of a hero of mine. Some of his buildings haunt me. In a good way. I admire original voices in all artforms, people who have studied the tradition of their art but then they take it to places it has never been before. Frank Gehry's architecture does that to the extent that some of his buildings look impossible to me. Yet still their landscapes adopt them, claim them, embrace them, so that soon you cannot imagine those landscapes without those buildings.

I read a newspaper article about his Fondation Louis Vuitton art museum in Paris that opened in October 2014. The article quotes the architect about the building's design:

> You have to keep that sketch quality. Anything overtly fin-
> ished is static to me. It's not living... We made the building
> ephemeral like a big sculpture. It's floating and changing. It
> looks like it's growing. No matter what you do with a normal
> building, it's a static building. This one isn't.

The author of the article, Jay Merrick of *The Independent*, observed:

> Each structural element of the Fondation Louis Vuitton, and
> every piece of glass, is a one-off... Frank Gehry has produced
> a building that comes together as if it's also just about to fall
> apart...

And then I asked myself an interesting question: "Do you suppose Frank Gehry ever watched beavers?"

☉☉☉

The top of the bank is thirty feet above the river. The bank is well wooded with oak, sycamore, willow, ash, birch, but especially oak. The densely mixed understorey is much cor-rupted by copious stands of Japanese knotweed, except that here and there it looks as if it has been trampled underfoot passing posse of buffalo, but this being Perthshire and the edge of the Highlands of Scotland, there has to be another explanation. There is.

There is a clue fifty yards away on the opposite bank. It is a long, low barrier of apparently randomly collected sticks, but with a flattish top and regularly sloping sides; imagine that someone had scraped the top off a bar of Toblerone. It seals off a narrow pool of still water from the mainstream. And the downstream end of the barrier is bedecked with a giant bouquet of knotweed leaves and a handful of late purplish-pink blooms. It is late October and the leaves are yellow and pale green and almost orange. They catch the sun and the breeze and my eye, and they make a fine show. But there is no knotweed growing on the far shore. That exuberant bouquet has been gathered from over here, carted down the bank (too steep and sodden and fankled by standing and leaning and fallen trees to be comfortably negotiated by me, but then I don't have four legs and a low centre of gravity), and ferried across the river through the boisterous, throaty, midstream current, which begins to gather momentum here for a series of rapids through a right-angle bend 200 yards away. It takes a seriously accomplished swimmer to cross that current, not to mention that such a swimmer must have had its "hands" and "arms" and possibly its mouth full of knotweed stems and abundant foliage at the time.

The dam-building, knotweed-toting swimmer has also been busy on this side of the river. The view of the bottom of the bank from the top looks like – well, it *is* – a building site. What you see amid such quantities of chaotically strewn raw building materials is a Gehry-esque sketch from which you may or may not get the gist of what he has seen in his head, but from which architecture will flow neverthe-less. Liquid architecture. It's like jazz, you know.

The riverbank has newly acquired a small bay, perhaps twenty yards across at the mouth and tapering to half that towards the foot of the steep bank. There had been land there, but it has been wantonly drowned. A single orangey, bluntly pointed face stares up at me from this new water, the surfacing head of a river serpent whose long, skinny body is still submerged. Closer inspection reveals that the "head" is the lower end of a felled tree trunk, and its "face" was chiselled by a beaver, every square inch of the surface not just denuded of bark but also decorated by roughly oblong bite marks, all of it effected in the process of the fell-ing. The rest of the tree is clearly visible underwater. It has been manoeuvred from where it was felled on the original bank, and deposited in the new bay where it awaits its new purpose, always assuming the beavers remember it's there. Beavers don't always remember such things.

The riverbank has also acquired a canal that runs par-allel to the river, rather like a mill lade. The beavers made this by opening up a weakness in the bank's defences fifty yards upstream, and it is this canal that supplies water for the new bay. A stretch of the original bank still divides canal from river, but it has become a long, thin island with trees, though not as many trees as it used to have. What the bea-vers are doing now is extending the downstream end of that island with the beginnings of a dam. Several tall, slim trees stripped of twigs and vegetation and quite a lot of bark sprawl across the outer reaches of the bay. The canal water's natural tendency to flow through the bay and rejoin mainstream is slowly being stifled. There is now a gap only a couple of yards wide and it is slowly being closed from both sides. Soon there will be a horseshoe-shaped dam between the land and the end of that outer bank; soon the water will have nowhere to go but the new bay, which will simply deepen and spread and become a bigger pool; soon the landscape will transform and the beaver will have a new sanctuary and its buildings will look impossible to me, yet this landscape will adopt them, claim them, until I cannot imagine the landscape without them. And yes, they are producing a building that comes together as if it's also just about to come apart.

All that assumes that beavers will finish what they have started building. Beavers don't always finish such things. But if they do, and for as long as they work on it, what they are building is liquid architecture.

You have to keep that sketch quality, you see, because anything overtly finished is static. It's not living. And this beaver architecture *lives*. It is ephemeral like a sculpture. It's floating and changing. It looks like it's growing. No, it *is* growing. Every day it grows. Every day it hosts a changing exhibition of improvised wood sculpture. Take that sea serpent head: tomorrow it may have drowned or been forgotten completely, or its parent tree may have been rendered down into bite-size chunks and transported to the new lodge, or across the river to that whatever-it-is that waves its knotweed banner at me. The banner will be gone tomorrow, of course, because the beavers will have eaten it or stashed it in their larder. That's what it's for.

☉☉☉

Nature's architect, then, is back in my native landscape. A Scottish Government-sponsored trial reintroduction of Norwegian beavers under strictly controlled conditions has been carried out in Argyll by the unwieldy coalition of the Scottish Wildlife Trust (SWT) and the Royal Zoological Society of Scotland (RZSS, but also more popularly known as Edinburgh Zoo), the Forestry Commision which owns the land, and all of it monitored by the Scottish Government advisors at Scottish Natural Heritage (SNH). That coalition has already consumed six years and a widely reported £2million. The beaver reintroduction project has, however, been supplemented (not to say discomfited) by an "accidental" population in Perthshire and Angus, the catchment area of the Tay, which just happens to be the biggest catchment area of any British river, a network of waterways, lochs, lochans, woodlands and bogs that looks as if it might have been designed specifically to facilitate the dissemination of beavers and beaver architecture over the widest possible area in the shortest possible time. And lo, it has come to pass. As I write this, the official trial in Argyll has been completed for a year, and a report on its findings has been presented to the Scottish Government, whose ministers will have the final say.

As for the Tayside population, it flourishes, it travels far and wide in strictly uncontrolled conditions, building and relandscaping as it breaks new ground untrodden by beavers for the better part of 400 years. SNH staff have referred to them as "the illegal beavers" and (wait for it) "the wrong kind of beavers", apparently because the original acciden- tal escapees may have been sourced in Germany rather than Norway. They are, nevertheless, the prolifically successful fruits of a handful of escapees from a captive colony perhaps

or more ago, and quite possibly augmented by the
.ɹ of qualm-free individuals on the outer reaches of the
conservation movement who have trouble keeping in check
their frustrations at the plodding gait of bureaucracy, and
who know where to get what they are looking for. That's
the illegal part, but nature doesn't give a damn. Nature is
doing the legal beaver coalition's job for it, and essentially
fulfilling the ambitions of the official trial. Almost inevitably, there have been dark mutterings in high places about a
cull, but given the farcical badger cull in south-west England
that began in 2013 and its attendant public relations disaster
for the Westminster Government, I cherish the hope that
Scotland will rise to the occasion, accept that the Tayside
population has been a happy accident, and enshrine in law
the freedom of all Scotland's beavers and their right to roam
where they will, and to practise liquid architecture.

And speaking of architecture, Charles Rennie Mackintosh, arguably Scotland's most innovative and original architect, gave a paper to the Glasgow Architectural Association
in 1891. Its subject was Scottish Baronial Architecture but I
have appropriated its message to do my bidding:

*This is a subject dear to my heart and entwined among my
inmost thoughts and affections... the architecture of our own
country, just as much Scotch as we are ourselves, as indigenous
to our country as our wild flowers... From some recent buildings
which have been erected it is evident that the style is coming
back to life again, and I only hope that it will not be strangled
in its infancy by indiscriminate and unsympathetic people...*

And then I asked myself another interesting question: "Do you
suppose Charles Rennie Mackintosh ever watched beavers?"

Chapter 2

FOOTPRINTS ON
A RIVERBANK

THE TELEPHONE RANG at an oddly early hour of the day. I answered through a mouthful of porridge, honey and banana, which is good for me but not for telephone communication. A furtive male voice as odd as the hour said:

"It's about the beavers."

"What beavers?" I asked.

"The beavers in *The Courier*..."

☻☻☻

I have a column every Tuesday in *The Courier*, the Dundee-based morning newspaper that serves east central Scotland, and the newspaper that started me off on a lifetime of working with words in one way or another at the age of sixteen. The morning of the telephone call was a Tuesday. My column that day was about beavers. The caller had been up very early to have read his *Courier* by breakfast time, but by

learned that beaver nuts are accustomed to keep-
unorthodox hours, because they keep beaver hours.

Before my column appeared there had been news stories
about an unknown number of beavers that had mysteriously
turned up on the pond of a trout hatchery in Perthshire.
One of the theories – the most plausible of many – was that
they had escaped from a wildlife park. A few days before the
phone call, the Scottish Government advisors, SNH, had
issued a confident statement to the effect that all but one
of the beavers had been captured. The gist of my column
was that I had what I thought was good reason to harbour a
smattering of doubt.

At the time, that coalition of SNH, Forestry Commission,
Scottish Wildlife Trust and Edinburgh Zoo had just
embarked on the process to establish a trial release of beavers at Knapdale in Argyll. In their own eyes, these were
"the legal beavers". The cost of the trial was a source of
some public disquiet, and some of the more restless natives
were talking out loud about direct action. One conversation
that reached my ears in a roundabout fashion seemed to
suggest it would be very easy to do the job of beaver reintroduction for nothing at all. To say I was intrigued would
be to undersell my level of interest.

Permit me to leave that early morning phone call on
hold for a moment, for it was not the first covert call I had
received on the subject of beavers. In the mid-1990s I had
made two radio programmes for the BBC about Scotland's
missing mammals and how they might be reintroduced.
"Rewilding" was in the air even then, only it wasn't fash-
ionable yet and those of us who were involved didn't call
it that, and nor was it anything like as far up the political

agenda, or even the conservation agenda for that matter. Among the people I interviewed was an old friend, Don MacCaskill, a retired (and very untypical) chief forester with the Forestry Commission. Don was a fine naturalist and an award-winning wildlife photographer who was constantly at odds with his professional employers because of his belief that the Commission of all people should grow showpiece forests that harboured indigenous wildlife as well as alien trees. The walls of his darkroom were plastered with photographs of wolves.

We did the interview on the boggy, willowy shore of Lochan Buidhe, which lies not far from the Perthshire village of Strathyre among the southernmost mountains and forests of the Scottish Highlands, and where Don lived. The location was significant for two reasons. One is that it also lies at the heart of what I have long regarded as my nature writer's territory, the landscape where I have done so much of my fieldwork that has fed into so much of my writing. It was also hand-picked by Don MacCaskill for a very specific purpose: it was here that he thought Scotland should reintroduce beavers.

"Why here?" I asked him. I had expected a thoughtful appraisal of the habitat and the possibilities of expanding the beaver population by way of a network of rivers, burns, flood plains, lochs and lochans. But he had already taken that as a given and moved on remarkably far beyond such basic considerations, and instead he replied at once:

"Because I had a dream about it. They were here."

With us was a Frenchman who had supervised beaver reintroductions in France, and he was smiling broadly, beaming approval at Don. I asked why.

"Because this is so like the region where we released our beavers in France," he said. Then he paused with his arms outstretched and looked around Lochan Buidhe, the nearby river, the fringing birches, alders, aspens and willows, the forests and mountains beyond, then he said:

"*SO* like it!"

Some years later I had recalled that conversation in my *Courier* column, and the next day I had the first of those two phone calls. The caller revealed nothing about himself but told me only that his attention had been drawn to my article. In summary the message that followed was this:

"Would you like some beavers? I know where to get some and they will cost you nothing. We can slip them into your lochan at night."

Nature conservation works at many levels, and some of them are as far underground as badgers. I imagined the caller as a skinny, spiv-like cratur with long, oily hair that fell over his face, and wearing a worn-out, third-hand Barbour jacket he had found wet and abandoned on a fencepost, and I was mildly concerned that he had my phone number. I declined his offer politely, explaining that at the time my writing base was about two miles from the lochan, that my particular philosophies in the matter of nature conservation were well known in the area, and that the local police would be knocking at my door before the first beavers had stripped their first yard of willow bark. Don had died by this time, so I would have been promoted from being deputy prime suspect to prime suspect. Yet even as I ended the call I cannot deny how deliciously tempting the proposition was. Beavers at Lochan Buidhe, the middle of my working territory. It would be perfect. I also cannot deny that it was

Don's revelation of his dream that turned my head towards the first faltering steps on a journey that culminates with this book.

All of which brings me back to that second phone call at an oddly early hour of the day, the one I answered through a mouthful of porridge, honey and banana, and on the same day that I had just expressed in print my doubts that there was only one "illegal" beaver at large, and the voice on the phone said:

"It's about the beavers."

The voice was oddly muffled, not muffled in the way that a mouthful of porridge, banana and honey muffles, but rather – if I had a suspicious mind, or a flair for the dramatic, or a taste for Ian Rankin novels (no, no and yes) – rather as if it might be muffled by a hanky over the mouthpiece. The unlikely component of that theory was that someone who made such a call would own a hanky, or, for that matter, a phone with a mouthpiece. So:

"It's about the beavers," he said.

"What beavers?"

"The beavers in *The Courier*."

"What about them?"

"There's *nine*."

And then the phone went dead.

◉◉◉

Don MacCaskill lived closer to the land, to nature, than most people. He was born at Kilmartin in Argyll, arguably Scotland's richest treasury of humankind's earliest marks on the face of the land – great burial cairns, stone circles, carved

13

stones, some with mysterious inscriptions. And dream, or something like it, shaped the minds and informed the knowledge of the mark-makers. Don may well have felt that nature itself had entrusted him with the knowledge that stemmed from the dream: *This is a fit place for beavers, restore them here.*

This is not as fanciful as it may sound. In fact, to me it sounds positively familiar. For I too have known moments of rare privilege, not in a dream admittedly, but certainly in nature's company, when it has felt as if I was being handed a rare insight, and I have written these down in terms that suggested nature was tapping me on the shoulder to command my closest attention. Don's dream sounds to me like one such moment. After he died, my time spent around the lochan acquired an added edge, for I thought I might keep an eye on his dream. All I had to go on was the Frenchman's beaming approval and outstretched arms, the little I already knew about what constitutes ideal beaver habitat, and – crucially – Don's calm faith in his own dream. I have never considered that these were inadequate justification for my irregular scrutiny of the lochan's waters at dusk, or its fluctuating shores as the river rises and falls and sometimes takes leave of its senses and floods a square mile of the glen; or my scrutiny of its landscape setting for unfamiliar footprints and trees that look as if they have been expertly felled with a chisel. The professional naturalist or the biologist may well scoff at my methods, but (as I am in danger of repeating too often) I am a nature writer, not a naturalist, and sometimes I dance to a different tune. I learned that tune from a woman from the English-Welsh border country called Margiad Evans, whose book, *Autobiography* (Arthur Baker, 1943), is as exquisite a piece of nature writing as you will ever find.

If you want to write with absolute truth and with the ease of a natural function, write from your eyes and ears, and your touch, in the very now where you find yourself alive, wherever it may be. Carry your paper and book with you and conceal yourself in the fields. Watch and be in what you see or in what you feel in your brain. There is no substitute even in divine inspiration for the touch of the moment, the touch of the day-light on the dream.

And here's the thing: a dozen years after he died in May 2000, Don MacCaskill's dream came true. A solitary, wandering beaver turned up at Lochan Buidhe, Loch Lubnaig and the River Leny, which drains the loch to the south, and there it was filmed by a canoeist. It did not stay, but beavers from Tayside have since begun to prospect in the Trossachs among the lochs and rivers a little further to the north, and I think they will settle there sooner rather than later. It turns out there was no need for early morning phone calls and hankies. The "illegal beavers" found their own way here all by themselves and made daylight of Don's dream.

◉◉◉

Every new thing has to begin somewhere, and this is where it began for me, one May Day, sitting by a quiet bend of a river near Callander about ten miles from Don's lochan, and watching the sun dip down behind the long south ridge of Ben Ledi. Twenty yards away over there and almost but not quite against the far bank, there was a strange arrangement of sticks. They lay in a compacted mass about twenty feet long, slightly higher at one end than the other and rising about three feet above the surface at its highest point. The higher end was also broader than the lower end, so it had something of the air of a surfacing whale. I was here two weeks before, and it was not. In the interim, the weather had been calm: this was not an edifice heaped up by a spate.

The first thing that occurred to me was that it was a pile of sticks on a river, so why did it not float away downstream, breaking up as it went, for although the river is relatively calm here, the midstream current is brisk? But not one stick so much as trembled. There was – there could be – only one explanation. Something was holding them in place. This was not a pile of sticks. This was architecture. This had been *built*. I inspected it in my binoculars. There were hundreds and hundreds of sticks. I had never seen a beaver dam or a beaver lodge before. Films and photographs, yes, but not an actual dam, an actual lodge, the actual pile of the actual sticks, sticks welded together into the actual pile none-too-prettily with actual mud. (The beaver, after all, invented reinforced concrete a million years ago.)

"I am not even sure this is what I think it is," I reassured myself. But nothing else explained its presence, and besides, there had been rumours, and there was that canoeist's video clip on YouTube reputedly filmed not far from here. I knew there were beavers in Argyll, I knew there were beavers widely scattered across the Tay catchment, but both of these were improbably distant from Callander. Even the nearest beavers I had heard of were on the far side of the very substantial watershed that divides the Tay's catchment area from the Forth's. Yet I had heard straws in the wind, those first hints that they had reached the wider Trossachs area just a few miles further north. What I had not bargained for was this, this pile of sticks not quite on my doorstep but encouragingly close enough. I was duly encouraged.

So I did what I always do when I begin something new. I choose my landscape. I go to where I think I should be, and sit down with binoculars, a camera (just in case: I am no-one's idea of a wildlife photographer), and more importantly by far, with a notebook and enough pens and pencils to stock a small shop. For me, the point of a working life in nature's company is, and always has been, to write it down. So I sit somewhere like this, not hidden, but discreet, as if I were a piece of landscape. And there I would wait in the dusk and prepare to get cold.

The sun had gone. The river muttered to itself in midstream, deep-throated and guttural, Old Man River. On the outside of the bend, which is where I was sitting, there was a scattering of small rocks, and here the river ruffled into wavelets and chattered and laughed and sang in a higher, playful pitch, Young Girl River. Nothing else moved. There were two patches of old snow just below the summit of the mountain.

A dog barked, a mildly irritating thin yap. "Whatever happened to wolf howls?" I asked the fading evening light.

A jackdaw flew west alone, talking to itself.

I had no idea what might happen next, so I expected nothing at all. I harboured no ambition beyond being a part of the landscape. If anything cared to show up and share it with me, well, I was here, ready to write it down, not to interfere, not to affect or manipulate, not to ring it or tag it or collar it or fit it with a transmitter (for such things are the essential tools of the trade of the 21st-century naturalist), none of these things, but rather I would simply watch, learn and write it down. That's what I do.

But nothing did happen at that pile of sticks, neither that evening, nor as far as I can tell, ever since, except that it was inundated by an autumn flood and when the waters fell it was still there, largely unscathed. If it was beaver's work (and I am as sure as I can be that it was), it was never more than temporary architecture, a dam to protect a lodge it had simply dug out of the bank, and had probably already been abandoned by the time I first saw it. It's still there, sagging quietly under the insistent tug of the water, slightly less intact. And two months after I found it and pronounced it abandoned, I found an alder tree about a quarter of a mile downstream and which was temporarily wading in floodwater, and it was as clear as daylight that a beaver had removed a yard of bark from the trunk. There is also this. Beavers (I now know) will return to old dams and restore them or rebuild them, sometimes many years after they have been abandoned. Good architecture is more or less immortal. Think brochs.

◉◉◉

So often, when something new begins, it does so with a false alarm, with a testing of the waters, with false trails and U-turns, with uncertainty, with doubt. One thing I was certain of: if I was going to write about beavers, sitting beside rivers was going to be a recurring theme. In a little backwater of the River Earn in Perthshire, a tiny burn a yard wide has cut a deep groove in a patch of lush, wet woodland. One of its banks tops a muddy slope and a muddier shelf just above the water, all of it overhung by tall grass, bracken and thick shrubbery that swarm around the trunks of big trees. This was a new piece of terrain for me, and it always takes me time to accustom to a new place. In the absence of the kind of familiarity, even intimacy, that I rely on within my working territory, it is necessary to walk slower, to be still more often and for longer, to listen harder and look harder.

In this mood of heightened sensitivity to my surroundings I found a yard of foliage above the burn that hung down in a strange way, so that I gave it a second look. If you were to ask me now what was strange about it, I couldn't tell you with any kind of conviction. I could tell you, however, that because I was looking for something and had not found it yet, I was straw-clutching, a circumstance with which I am not unfamiliar on those days in new surroundings when nature appears to be unwilling to co-operate with me. Even after all these years, I still feel as if I have to prove myself to nature all over again, to establish my credentials, to reveal the benevolent nature of my presence. In any case, I have already asked myself the same question. Why did I give that hanging greenery a second glance? Only this: nowhere else along the top of the bank did the vegetation hang down. There was no sign of animal tracks through the vegetation to suggest it might have been flattened, and it had not rained there for weeks, so neither rain nor the river had flattened it either. What did occur to me was that it looked as if an animal might have pulled it down from below, and that was wishful thinking on the part of the straw-clutcher.

High summer is the worst possible time to explore somewhere like this without a boat, preferably a canoe, and not only do I not have a canoe, even if I did I don't know how to handle one. What I do have is a pair of wellies, a good stick, and an incurable addiction to the off-piste in nature. I had been driving a back road near the river when I saw something unusual across by the far bank, and in that instant I recognised what could become the beginnings of a pattern. It snagged in a corner of one eye, and no matter that the eye in question should really have been looking at the road. But the road had been empty for miles, I was driving slowly, and having seen what I saw, I stopped by a very old and very overgrown iron gate.

The place spoke to me, the way places sometimes do. It said, "This is the land that time forgot," or Housmanesque lines to that effect. Something like a track appeared to have connected gate and riverbank long ago and for a purpose that was no longer apparent, perhaps a ford, perhaps a wooden landing stage that had succumbed to a century of winter spates. But it was all so long ago that nature had since reclaimed it and high summer had embellished the reclamation with dense and head-high shrubs, waist-high nettles, and improbably deep holes and ruts full of dark green water. I needed something more negotiable, yet it was clear from the forlorn aspect of gate and track that the place was undisturbed, and undisturbed was exactly what I wanted. I back-tracked and wandered along a stretch of road looking for another way down to the river.

It was then and there that I saw the burn, its yard-high bank with a yard of curiously displaced vegetation on top, curiously, that is, if your mind runs to the conspiracy theories of the incurable straw-clutcher. And then I saw that the muddy shelf at the bottom of the bank, an inch or two above the surface of the burn where it entered the river, was patterned with footprints. Some of them (I saw at once through the binoculars from about thirty yards away) were roe deer, but the others were not, and it was because of these other footprints that I decided to tackle the steep slope between road and river head-on, rather than look for anywhere else more negotiable.

But at that moment, just before I tackled the slope, I realised I could see again that particular unusual something on the far side of the river that had persuaded me to find somewhere to stop in the first place. Two thin branches, each about fifteen feet long and with foliage at one end, lay half in and half out of the shallows. I could see clearly now that the branches had been newly severed from their parent tree, and that there were several long, thin, white strips where the bark had been removed, so now I had two good reasons to find a way down. I slithered down the embankment towards the burn, waded across dry-shod in wellies until I was right beside the muddy shelf with the footprints. There were two perfect prints that looked like little hands, and overlapping them slightly from behind were two more, larger and deeper-embedded, and apparently webbed because they looked like five toes bursting out of a sock. This was what I had hoped to find in my wildest, straw-clutching dreams. All this – the oddly dragged-down vegetation, the two half-stripped branches, and the footprints (rear feet webbed, front feet unwebbed) – amount to the unmistakeable signature of beavers, which is what I had been looking for all along. Beavers had been feeding here very recently. That thought rather stopped me in my own tracks (wellied, size nine-and-a-half, unwebbed).

This was like finding a cave painting or a yard of Ogham

script carved into a stone slab in Kilmartin, runes executed by the mark-makers of history. Before this square yard of mud was imprinted, a few hundred years lay between me and the last time *Castor fiber* was a familiar presence on Scottish rivers and lochs. And it was on this particular river that this particular representative of *Homo sapiens* picked up the spoor again for the first time. The thing that I marvel at now, a year after the event, a year in which I have grown accustomed to the landscape-transforming imprint of the beaver as lumberjack, engineer and architect, is that my first confirmation of its presence was a set of footprints in a square yard of mud, which I only found because I backed a hunch.

It was a bright, sunny afternoon, and beavers are mostly dawn and dusk creatures, but before you start staking out a likely-looking piece of river at unsociable hours, it is good to be plugged into the kind of grapevine that mutters reliable intelligence in your ear from time to time, and to check out for yourself that the necessary clues are in place. Then you can really go to work.

The beaver is back in our midst. The numbers are still small, but their progress has been resolute and impressive. Nature has also gone to work, and doesn't much care whether the powers of the land think the beavers are legal or illegal, nor whether they are electronically tagged, nor whether they are equipped with no more sophisticated communication equipment than the means they already use for communicating with other beavers, and have been using for a few million years. Nature also knows that there is no such thing as an illegal beaver.

Chapter 3

THE RIVER DIARIES: I

A SHOT ACROSS the bows may be appropriate at this point. It was offered by Horace T. Martin in his introduction to a Canadian book titled *Castorologia* and published nearly 125 years ago.

> *A traditional knowledge of the beaver is the birthright of every Canadian; yet, as in most cases where tradition alone is relied on, this knowledge is chiefly remarkable for its divergence from the facts. As the acorn, falling on favourable soil, sends forth the slender shoot, which time and circumstance may model into a grotesque fetish for minds ignorant, or forgetful of the simplicity of the origin, so the facts of science, if nurtured by tradition, soon lose shape, and multitudes venerate the fabulous stories of dragon or beaver, with total disregard to outraged reason.*

You have been warned. So you are invited to substitute the word Scot or Briton as you see fit in the first line (for the sentiment is no less true here), and harken to the voices of outraged reason.

⊚⊚⊚

A shudder went through the surface of the water. No sooner have I let the word "shudder" loose on the page than I want to haul it in, but just a little. I want to tone it down. If I was a watercolourist (how often, keeping nature company, I wish that I could paint) I would mix in half a brushful more water to lighten the shade, just enough to be noticeable on the paper to discerning eyes, not enough to devalue the deeper shade I chose in the first place. The change would reflect the sudden awareness, the new knowledge that was not there a moment ago when I had thought that what the surface of the water did was shudder.

Another problem is that right then, even as I tried to introduce a shimmering quality into the shudder – a shummer? a shidder? – right then, I didn't know what was going on. There is no question that something was untoward near the far bank, for the water that was so tranquil and almost still, almost at a standstill (a conspicuously unusual quality in a Highland river, even one as borderline Highland as this one)… that water was suddenly beset by shallow corrugations advancing *upstream*, and in the pre-dawn gloom energised by a low and fitful mist, there was an eerie cast to the troubled surface.

Over here on my side of the river, the current was brisk without breaking into a canter or making waves; yet I could see the upstream disturbance (which I would rank somewhere between wavelets and ripples – waveles? ripplets?), see it sprawl across the current in ever diminishing troughs until they reached that critical point in midstream where the current was too strong and their outer edges were bent back and dissolved. Yet something must have generated the original disturbance, given it impetus and then sustenance so that it flowed upstream in the quiet waters of the far bank. And in that first moment of my awareness of whatever it was, I was startled into a stillness of my own, thinking that the surface of the water had just shuddered. Then I had thought it less than a shudder, but not much less, the confusion engendered by bafflement.

The thing to do now, I told myself, is nothing at all. I don't mind telling you, I am particularly good at doing nothing at all. It is a gift that serves a nature writer well, and although I can take no credit for it — for I appear to have been born with it — I am at pains to nurture it by means of constant practice. This was a moment for utter stillness, but perhaps a more watchful stillness than the one that just caught me out when I thought the water shuddered. So I did nothing at all. And then, as if by way of a reward, a tree started to swim towards me.

It had to be swimming rather than just floating because it too was moving upstream against the tug of the current. It was not a whole tree but a substantial part of a small one, and with foliage trailing out into midstream. This singularly accomplished tree was, I now realised, the cause of the shudder, the wavelets, the ripples, the bafflement. This tree must be inspected.

It stood to reason that it had to have a source of propulsion, otherwise it would turn in its own length and float away downstream, where, a few miles further on it would enter the Tay, pick up speed, blur past Newburgh, next stop Dundee, and by nightfall or daybreak it would be somewhere out by the Bell Rock lighthouse. Instead of that it had covered twenty upstream yards in about half a minute, and as it drew level with me I realised that a curious bulge that deformed the otherwise straight and slender trunk was the top half of the head of a swimming beaver. At this point in my constantly evolving relationship with nature, I had never seen the head of a swimming wild beaver before, with or without a tree attached. So this, in the greater scheme of things, was quite a moment.

⊙⊙⊙

The simple way in was to listen. Thirty years of writing about nature for a living have taught me how to put my ear to the ground, especially when there is something quite new afoot; or, as in the beavers' case, something very old that has accepted an invitation to recolonise its historic heartlands. That invitation came in the first place by way of the far-sighted zeal of two singularly untypical Scottish landowners, Sir John Lister-Kaye and Paul Ramsay. I have learned a great deal about beavers from both of them (see chapters 8 and 10), but still I needed to establish my own study area for myself. The animal becomes more alive, more vivid, more captivating in the eye of the beholder and in his imagination, and the insights are deeper and truer, when they are gathered in silence and solitude. So I put my ear to the ground and listened, I found my river, I went to work, and I found footprints in a square yard of mud.

Then I saw a tree standing right at the water's edge with a bite out of it. People who live near beavers are all too aware of them even though they rarely see them. Like badgers, beavers are creatures of the dawn and the dusk and the hours in between, but they yield to the long daylight evenings of spring and summer and show themselves willingly enough then in sunsets and gloamings. And (as I had yet to learn) also like badgers, they occasionally confound all the received wisdom and emerge in the quietest hours of the quietest places in broad daylight. So if you do have beavers nearby, mostly you don't see them, but the trees bear the scars, or they disappear from their accustomed stances and suddenly water flows where they were accustomed to stand. And here was my first tree with a bite out of it. From a distance of about twenty yards it looked like the kind of bite such as a tree-munching dinosaur might have inflicted, but I have no idea if such a thing ever existed. Dinosaurs are not really my field.

My response to my first sight of my first beaver-bitten tree in Scotland was to freeze, then to check wind direction, then to look around for somewhere to sit and watch it. At this stage, I had no idea whether beavers rely mainly on scent, sight or hearing, or any combination of the three, to detect intruders. The place was sheltered enough for wind not to be much of a problem. Nearby was an unbitten tree higher up the bank from the bitten tree, and I reckoned that in my tree-coloured clothes I could sit with my back to it and watch. There was no animal in sight at that moment, and given that it was still an hour before sunset on a late autumn's late afternoon, I reasoned that I had time on my side.

Hides are anathema to me. My preferred way of working is to be discreet, but also to try to be a visible part of the landscape rather than a hidden one. I *want* nature to see me, and over time to recognise me for what I am, an essentially benevolent presence rather than a threat. But here, in the first hours of this adventure into Beaverland, I was aware that I had become very tense, when I wanted to be calm. Tension transmits itself to many animals, especially if it is cloaked in the guise of the creature responsible for past sins against the animal in question, which in this case would have to include extermination. Troubled by this state of mind, I moved awkwardly, executed a sidestep the better to focus my binoculars on the tree, the bank and the river. My camera, which I had slung over one shoulder, slipped off in response to the ill-judged movement and bounced against tree bark. It was not a loud sound, and I caught the strap at once and froze because instinct commanded me to freeze. It was a well-judged command. There was an immediate muffled explosion behind the bitten tree, a loose branch clattered against the roots and bounced into shallow water. A hefty, dark grey-brown and bluntly rounded shape leapt into the river, followed at once by a startlingly loud splash from what I now know to be a beaver tail slapping the surface of the water as it dives. In the process, it alerted every beaver and every other creature along several hundred yards of river to my presence.

Beaver tails are weird. They are like no tails you have ever seen before. They are wide and more or less flat, and tapering only slightly at the far end, as if they had been put through a mangle. If you don't know what a mangle is, your grandmother may be able to help you. Beaver tails are not covered with fur like regular furry mammal tails, but with hard scales. In the wrong kind of light, it can look as if the beaver is towing a fish. At first, because of the remarkable nature of the encounter, I presumed that its primary function was to sound the alarm, an unambiguous "bugger off" to me and all my ilk.

◎◎◎

More on beaver tails. This from *Castorologia*:

> *It is nearly flat, broad and straight, and covered with horny scales of a lustrous black. Its principal uses are to elevate or depress the head while swimming, to turn the body and vary its direction, and to assist the animal in diving. When alarmed in his pond, particularly at night, he immediately dives, in doing which the posterior part of his body is thrown out of the water, and as he descends head foremost, the tail is brought down upon the surface of the water with a heavy stroke, and deep below it with a plunge.*
>
> *It is capable of diagonal movement from side to side… and also of assuming a near vertical position.*
>
> *He is able to turn his tail under him and to sit on it, or to use it extended behind him as a prop when sitting on his hind feet.*

So it's a rudder and a picnic stool too. Moral: don't jump to early conclusions. Second moral: don't underestimate anything about beavers.

◎◎◎

It is not quite dark inside. There is a hole in the roof, a ventilation shaft dug into the top of the riverbank. It is just possible it can also serve as an emergency exit/entrance. A little late light from the sunset's afterglow finds its way down through the dark earth that has been tunnelled and hollowed out into a random complex of chambers. Is everything a beaver does both random and complex?

There is an entrance tunnel that leads to a doorway that opens into the lower depths of four feet of water. That depth of water so close to the riverbank where recently there were only inches-deep shallows has been achieved by a beaver-built canal that leads to a beaver-dammed pool surrounded by river-fringing trees, some of which the beavers have felled to make the dam, some of which may be intended for the lodge beyond the far bank. Inside the lodge a beaver marks the fading of the light and stirs. It is time to be up-and-doing. It seems already that almost every waking moment of every beaver that ever lived is filled with time to be up-and-doing.

The roused beaver stands, at which point it looks like a large grey-brown tea cosy. If you don't know what a tea cosy is, see "mangle" (above). It shuffles clumsily forward from chamber to tunnel, slithers from dry earth to wet mud to water, at which point it transforms, becomes as lithe as an otter. Beavers, you learn quickly, are shape shifters. A beaver in the water looks a little like an otter in the water. An otter out of water still looks like an otter. A beaver out of water looks like a tea cosy. A tea cosy that tows a fish.

It swims down to the bottom then shoots upwards and forwards with a thrust of its webbed hind feet, breaks the surface without a splash, without a sound, and there it lies quite still. And there it eyes the overworld and its sky from the meniscus of a waterworld so even-tempered, so agreeably and elementally attuned, that it seems almost sacrilegious to call it architecture. Yet architecture is precisely what it is.

The wide oasis of purposefully diverted river water, where the beaver lies like a log in a daydream, has been designed, engineered, project-managed, and built; built with beaver hands and feet, and beaver teeth. Likewise the dam that holds it all in place, likewise the lodge with its air vent, watergate, and seemingly haphazard network of living spaces and larders. Likewise the canals that radiate from riverbank into woodland: these offer fast escape from shoreline wood to the safety of deep water, escape from such historic threats as wolves and wolverines that must still throng the race memory of 21st-century Scottish beavers. Even though neither wolf nor wolverine torment the landscape of Perthshire's Highland Edge, both species are still active in Norway and other parts of mainland Europe, and even though there is no historic evidence of wolverine here and the wolf has been gone for at least 200 years, who knows when the people will reintroduce the wolves? And who is going to tell the beavers when it happens? So they listen to their instincts and they dig canals, and for now they are unlikely to encounter any more troublesome threat than a guided group of tourists with cameras, or someone's dog, or a solitary nature writer caught unawares and startled into a noisy false step.

So a beaver surfaced from gloom into daylight and lay where it surfaced, floating and idle, at least in the sense that it was still, but even in stillness it was up and doing, its eyes, ears and nose examining the evening, assessing its possibilities. And after my first dozen visits, I saw it surface and grow still from a long and painstaking stillness of my own, and despite appearances, I too was up and doing.

⊙⊙⊙

The river rises in the west from sources high on the eastmost slopes of the Balquhidder hills and high on both sides of Glen Ogle above Lochearnhead. All this, and a myriad other hill burns, feed Loch Earn, from the east end of which the River Earn flows fully formed and heads for a union with the Tay about forty miles away. Its character, where I have come to watch beavers, is unquestionably Highland. It boils over rapids and tumbles over shallow falls. It grows ominous and bloated when winter rains and mountain snow-melt conspire. Its banks are wooded here, mostly with oak and birch and alder and willow, any or all of which grow right down to the water's edge, and every winter, the spates tug a few of them loose from the bank and cart most of them off into a fast oblivion. But some snag against banks or rocks or shingle, and some of these the beavers first liberate then convert into their constituent parts of food and building material.

Over the next few eastward and downstream miles the river evolves subtly into the broadening and increasingly agricultural girth of Strathearn. The hills lower and fall back, the river slows and dawdles, until by the time it meets the Tay, it has become a Lowland meanderer. It occurred to me one quiet afternoon sitting here that its life is the diametric opposite of my own, for I began as a Lowland meanderer and evolved westwards and northwards into a would-be Highlander daring mountain snows and mountain waters, wandering among eagles, hankering after wolves. It intrigues me that in a phase of my life when I have been thinking more about who I am and the importance to me of my native landscape, even my work has subconsciously reoriented from the sunset seductions of the West to this river at my feet that shakes itself free of every night and heads for the sunrise coast I still think of as home. I have always considered that the chart of my life lay between the compass points of North and West. Right now I wonder if this beaver river, tugging eastwards at my boots like an impatient collie, is trying to tell me something.

And then it occurred to me that the beaver is doing more or less the opposite. It escaped into Lowland waters, and over much of its range in mainland Europe it is a Lowlander too. Even the official trial in Argyll was established in woodland at sea level. Here too, it has set up an outpost on the Highland Edge, a nicely wooded halfway house between long, slow Lowland waters and short, fast Highland waters. But of course it has gone beyond now, made its way into the Trossachs, and from east Perthshire to west Perthshire, heading toward Rannoch and Glen Dochart, which are uncompromisingly Highland. We – the beaver and I – have crossed paths at a time when we are both re-evaluating our place on the map. I feel good about that.

◎◎◎

In the course of writing a handful of books about a single species – *Waters of the Wild Swan* (Jonathan Cape, 1992), *Badgers on the Highland Edge* (Jonathan Cape, 1994), *The Last Wolf* (Birlinn, 2010) and *The Eagle's Way* (Saraband, 2014) – there is a common element to the fieldwork that underpinned them all. It is that by concentrating so determinedly on one creature you find yourself drawn into the lives of every other creature with which it interacts and every other creature with which it shares its landscape. Nothing in nature lives and works in complete isolation. It has proved no less true of beavers than it was with any of the others. It became quickly apparent during my first two visits to that stretch of the Earn I am now apt to refer to as Beaverland, that in addition to such essentially river-thirled creatures as herons, dippers, kingfishers, sandpipers, salmon, trout, and otters, I would also be seeing a lot of red kites. There is no relationship between the kite and beaver that I am aware of, and I am having difficulty contriving circumstances in which there could possibly be a relationship. But on my very first afternoon at the river, I heard a kite calling within the first half an hour (a cry like a happy buzzard, if your imagination can embrace such an image), a call that was answered at once by a second bird, and there they were briefly crossing the river about 100 feet up, vivid chestnut red in high summer sunlight, white underwing patches flaring like headlights as one of the birds banked.

I knew the kites were around long before I knew the beavers were here. They are regular visitors to the airspace over nearby St Fillans golf course where I have played for a dozen years now, and occasional predators in its wilder corners rarely explored by golfers (although if you are discomfited by a bad hook on the sixth…). They are often seen hunting over the high moor south of Comrie, and there is a winter roost not far off where they gather in the evening in spectacular numbers, performing a kind of ritual aerial ballet before settling down for the night. Just a few miles over the hills to the south is the red kite feeding station at Argaty near Doune which is open to the public, and which between 1996 and 2001 was one of the release sites for the red kite reintroduction programme. So I am no stranger to red kites. On purely aesthetic grounds I think them among the most beautiful fliers in the land, so I was happy to know I would have their company at least sporadically.

They have in common with beavers that they have been rendered extinct in Scotland and have been reintroduced. Extinction happened around the end of the 19th century, and around the end of the 18th in England. By the late 1970s, what was left of the entire UK population of red kites could be traced to a single female bird in Wales.

In Scotland, the culprits were the usual suspects – the Victorian sporting estates, the Victorian egg-collecting mania, and the Victorian enthusiasm for killing things to stuff and display in glass cases, having deprived the kite of its beauty and its powers of flight in the process, as though that were somehow an improvement on watching a living, breathing, beautiful one out-flying the wind across a hill-side. The disservices the Victorians rendered on nature across Britain and across Scotland in particular are unquantifiably many and various, and the attitudes that spawned those disservices are, alas, far from extinct. In the space of two weeks in April 2014, and over an area of two square miles of the Black Isle, north-east of Inverness, the carcases of sixteen birds of prey were found in woodland. They were four buzzards and twelve red kites. The Black Isle was the site of the first reintroduction in Scotland: ninety-three Swedish birds were reintroduced between 1989 and 1994. At the same time, a similar number of birds were reintroduced into the Chilterns in Buckinghamshire. Among egg-laying birds, the record of productivity is the same, but today's Chilterns breeding population is ten times that of the Black Isle. An awareness of killing on that scale is surely as deeply ingrained in the DNA of the reintroduced beaver as it is in the DNA of the red kite. That is the nature of the relationship between them.

In late summer when I first came to look for beavers on the river, it was mostly in the deep shadow of the well-wooded canopy, and only a narrow blue furrow of sky that followed the course of the river let in much light. But during the moments when my mind was not utterly preoccupied with the beavers it was there that I would see and hear a few fragmented seconds of kite flight. Their fleeting, shimmering sunlit dance reminded me incongruously of those glitterballs that used to hang from the ceiling of dance halls (not that I was a dancer, but I played in a few dance bands). At first, I was hooked on this heartland of the beavers' realm with its constantly evolving timber operations, the innumerable questions it posed, the infrequent glimpses of the critters themselves and the strangeness of them to my eyes. And the whole place has an introverting, captivating ambience that has never really diminished in my mind. I would drive in by the little access road above the river, park as discreetly as I could, inspect the various hives of activity within the space of a couple of hundred yards of riverbank, then sit and watch in one of three places I had considered would make good vantage points. And while I sat and waited for something to watch, I wrote in a notebook on my lap. And every now and then, the sky would call down to me through the gap in the canopy and my head would jerk up and there was a kite, or a pair of kites, mocking my willing incarceration in the shadows, while they flirted with wind and sunlight.

It was late autumn the day I first decided to explore upstream. There had been a hard frost and ice had nibbled at the shallows. The last bars of the anthem of the red deer rut still rumbled down the hillsides to the south. I walked out of the trees, followed the road west into much more open country, and within minutes I had turned to raise my binoculars to what sounded like more than one buzzard. High above the woods to the south of Beaverland there were three buzzards wheeling and side-slipping, diving and soaring, and the excited edge of their calls cut brightly through the clear, still air. And with them were four red kites, and their keener and giddier voices laced the melee with added confusion.

This tumbleweed of birds was drifting apparently involuntarily northwards. If I had stayed among the beaver trees, it would have passed directly overhead, and while I would have heard all of it and strained my frustrated eyes to piece together what was going on, I would have seen next to nothing. Out here, I had a box seat. The red kite is a slender, leisurely, airy waltzer in the macho world of raptors, and a buzzard is halfway to being an eagle, with many of the eagle's traits. I am guessing now, but this looked much more like stuff-strutting than dog-fighting, point-scoring perhaps, striving for status. The buzzards fell the way eagles fall (although if I am picky, with none of the eagle's gravitas and panache), and from time to time within the chaotic, haphazard nature of the group one would climb above the kites and attempt to dive down through them, and each time the kites split in four different directions, a last-gasp waft of guile so that the buzzard fell through the hole in the doughnut, as it were.

Gradually, I discerned a pattern. The kites seemed to be on the move from east to west but the buzzards seemed hell-bent on turning them north, or at least somewhere north of west. Each time the kites reunited after a buzzard dive they started west again, but one buzzard after another would mob them from above, below and behind, and the morass of birds would edge a little more towards the distant hills in the north. It was as if the kites were an immovable impediment strewn across the buzzards' road north and must be moved aside like a snowplough moves snow. How long this might have lasted is anyone's guess, but as it happens I foresaw its denouement a second or two before it actually happened. I was suddenly aware of another bird in the glasses much higher than the group I was watching. Then there was another one by its side, then another, then three more, then... so I shifted the focus of the binoculars onto this higher group and I saw that there was a second tier of birds, that there were about forty of them, that they were moving west, and that they were all red kites.

Suddenly the buzzards were drifting away to the north and the four lower kites amalgamated seamlessly into these higher squadrons and their majestic and suddenly silent bandwagon rolled on. Now, tell me, why anyone would want to poison such birds?

Chapter 4

THE BUTE EXPERIMENT

In a solitary pine wood near Rothesay on the Isle of Bute, a space of ground has been walled in so that the Beavers cannot escape, and through this Beaver's park runs a mountain stream. Left to themselves, they have quite altered the appearance of this stream...
  – James Edmund Harting, *British Animals Extinct Within Historic Times* (J.R. Osgood, Boston, 1880)

FIRST OF ALL, it was not a mountain stream in anything other than Harting's imagination. Bute has no mountain streams because it has no mountains, and for that matter, no land above about 450 feet. The burn in question rises from a moorland plateau that lies at around 300 feet. The island was once the vigorously beating heart of the historic county of Buteshire, an island fiefdom in the Firth of Clyde that also comprised Arran, Greater Cumbrae, Little Cumbrae, Inchmarnock, Holy Island and Pladda; county town Rothesay. The local government upheavals of 1974 made nonsense out of that coherence by creating a municipal lump known as Argyll and Bute and North Ayrshire. At a stroke, the island of Bute and Rothesay were relegated from the municipal mainstream to the back of beyond.

The ferry sails from Wemyss Bay with at least one passenger in a disgruntled frame of mind. To get there from where I live there was a forty-mile drive to the Erskine Bridge across the Clyde, the fag-end of a motorway, then a drive through Greenock in the rain, which only felt like it took twice as long as everything that had preceded it, and such was its sustained disagreeableness that my mood was still suffused by it when I began to register glimpses of the suddenly lightening and widening firth, after which I entered a one-sided street of Victorian villas with sea views, villas that mostly appear to be struggling with the realities of the 21st century. What survives of the built heritage of the Clyde coast's glory days is not surviving in any degree of comfort or wellbeing.

The ferry awaited. I bought tickets for myself and my car in a grey hut and for a smallish king's ransom. This quest for some kind of historical context in which to set this new beaver age had better be good, I warned myself. But I drove straight on, the ferry sailed almost at once, and I began to cheer up. I like ferries, and I like going to new islands, and for me, Bute was a new island. It had been a long time since breakfast. I thought "coffee and bacon roll" and headed for the café. The woman in front of me had had the same idea. Her request was met with a polite smile, and this:

"We ran out of bacon rolls a while ago."

She smiled determinedly back:

"And what time do you have to be up at to get a bacon roll on this particular luxury cruise liner?"

Another smile, easy, practised:

"Earlier than this."

The woman settled for coffee and cake.

Yet another smile, for me this time:

"Yes, sir?"

I meant to say that I was on the same roll-and-bacon mission as the woman but it came out as "roll and beaver".

A frown where the smile had been:

"Sorry, what did you say, sir?"

"A roll and butter."

"Sorry, sir, no rolls. No bacon."

Another smile:

"No beaver."

Rothesay in the rain reminded me of nothing so much as Wemyss Bay in the rain, which is to say that it looked as if, between ferries, there was nothing much to do and all day to do it in. But five miles to the south of Rothesay, the relevant "mountain stream" should be flowing under the road.

I followed the signs for Mount Stuart, ancestral home of the Marquesses of Bute, where I had an appointment with an archivist who had already done some rummaging through late 19th-century files on my behalf. And somewhere down the road and inland a burn had once flowed through a "Beaver's park", and had its appearance much altered as a result.

What I was looking for was this. The 3rd Marquess had acquired some beavers in the 1870s, by which time they had been extinct for between 200 and 300 years. I wanted to know why and where he got them. In particular, I was hoping to talk to the current Marquess of Bute – the 7th – for I had assumed that such a significant and trailblazing event (significant at least to a nature writer on a mission to give a historical context to this new age of beavers) might be something he was rather proud of and eager to expand on. As it turned out, I'll never know.

The 7th Marquess of Bute, John Crichton-Stuart, Earl of Windsor, Viscount Ayr, Lord Crichton of Sanquhar and Cumnock, Viscount Kingarth, Lord Mountstuart Cumbrae and Inchmarnock, Baron Cardiff and Viscount Mountjoy, was also the Earl of Dumfries until 1993, which is why, when he became a motor racing driver, he adopted the name of Johnny Dumfries. In that guise he had a notable head-turning duel with the young Ayrton Senna in Formula 3, which led to a triumphant British Formula 3 season that included ten victories. An ill-starred move to Formula 1 lasted only a year, but he found renewed success with sports cars, notably winning the Le Mans 24 Hours in 1988. But he does not live at Mount Stuart and he was certainly not available to talk about beavers. Instead, I had been alerted by very helpful staff to the possibilities of the Mount Stuart archives.

Meanwhile, the glorious bulk of Mount Stuart materialised through the trees. My Dundee Corporation prefab upbringing hardly befitted me to feel at ease in such an ambience, but the architectural glory of the place stirred undeniable excitement. Gothic revival was never my favourite architectural time zone, but this... this was something other, the inspired work of Sir Robert Rowand Anderson. The original Georgian Mount Stuart was built by Alexander McGill in 1716 but the central block was destroyed by fire in 1877, and it was this that Anderson re-envisioned. Work began in 1880, continued until 1912, and in the grand tradition of grand designs, is still not finished. I like to think Frank Gehry would have approved. This is not a static building either.

My appointment was in the office, which was discreetly

signposted well away from the main entrance. A quiet path led to a humble door, a humble corridor within, and a humble, windowless office where I announced myself and the nature of my mission. If there was a smaller room in the house, I didn't want to see it.

Someone came to meet me, spirited me away through featureless corridors and doors, as undistinguished and undecorated as mineshafts. Then one more door opened and I stepped out into the Marble Hall, eighty feet high, adorned with signs of the zodiac stained glass windows, and a ceiling of the night sky with the stars in their allotted portion of the heavens. I heard my own gasp, and so did my guide.

"It gets everybody that way the first time," she said.

Our destination was the library, which would have been daunting enough had I not come by way of the Marble Hall, and frankly, by now I was beyond daunting. Files had been laid out for me on a huge central table. They were all personal correspondence, some of it in the 3rd Marquess's hand, some of it written by his solicitor Francis Anderson, some by Charles Jamrach, whose headed notepaper read:

Charles Jamrach, Naturalist, London, Dealer in Birds, Waterfowls, Animals, Birdskins, Shells.

I had found the source of the Isle of Bute beavers.

But by the time the first of these letters was written, the Marquess's beaver adventure had already begun, and there was nothing at all to indicate what had prompted it, although it became clear from Mr Jamrach that the Marquess was amassing a menagerie of sorts, and the inhabitants of the beaver park were just a part of an enterprise that, apart from anything else, suggested that unimaginable wealth was funding it, just as it was shortly to fund the rebuilding of Mount Stuart.

On the 12th of July, 1874, John Patrick Crichton-Stuart, 3rd Marquess of Bute, wrote in a letter to Lady Bute (who was obviously furth of Mount Stuart at the time; they also had a London home):

> *There are two more beavers here, wretched little things – I suppose young ones out of a nest. They are still in their box… One dare not put them with the old ones who have been put in a horrible temper ever since the new ones were put in sight of them.*

So there are already old ones in situ, and both they, and the Marquess by the sound of it, are less than enamoured. A month later, he wrote in a second letter:

> *The beavers were turned out, but they hid themselves so successfully yesterday that we couldn't find them when we went to look – I believe they will do all right.*

They didn't. Their fate is unknown beyond a subsequent observation that they "failed". But in January of 1875, His Lordship acquired seven more, and these would do rather better.

A dozen years later, Charles Jamrach lets a small cat out of the bag in a letter to Mr Anderson, who seems to have been the go-between in the Marquess's dealings in this sphere of his many activities:

> *If His Lordship takes the 3 Canadian beavers off me the price will be £10 each… I only get a very slight profit out of them*

to sell them at £10 each, which is not worth to me the risk unless I can send them away at once.

So he is dealing in Canadian beavers, *Castor canadensis,* rather than the native Eurasian beaver, *Castor fiber,* and he sounds a bit like the worst kind of car salesman. There is a subsequent exchange about still more beavers, the highlights of which are as follows:

Sorry to hear the beaver you got from me has died. I shall be glad to hear from you if you wish these beavers to be sent.

And the next letter:

I am sorry I cd not succeed with the beavers. The captain was obstinate to the last and would not budge from £12. He has sailed for Germany taking the animals with him.

Only one side of this long correspondence survives, which does rather make me wonder about the nature of the request to Mr Jamrach that prompted the following:

I do not think it wd pay to have wallabies brought over from Australia especially as the expense and risk wd be too great… One pair has just arrived. I can let you have the female for £10 or the pair for £20.

And this:

I also have 2 small sloth bears about 6 months and 8 months… Would make capital pets at £10 each.

And this:

Please note I am sending you tomorrow by night mail via Glasgow two female Bennett's Wallabies.

And there the whole bizarre enterprise might have vanished without trace, leaving 21st-century scruples to dwell scornfully on such misguided horse-trading that first trapped then crated wild animals and despatched them off on interminable sea journeys to travel the world so that they arrived at last in the hands of some Charles Jamrach or other with a copious contacts book stuffed with the addresses of the great houses of the land.

But, luckily for posterity, the 3rd Marquess's beavers were entrusted into the care of one Joseph Stuart Black, who became Mount Stuart's Beaver Keeper. And he was interested enough and attentive enough and articulate enough, to write a brief account of the enterprise, and which was published in the *Journal of Forestry* probably in 1880, just a year before he died. And this is what he wrote:

In 1874, the Marquis of Bute having obtained four beavers, caused a space of from three to four acres in extent to be enclosed in the wood between Meikle Kilchattan and Drumreach, and placed them there. These not succeeding, his lordship, on 6th January, 1875, obtained seven others. Of these, four succeeded so well that in 1878 I was certain of sixteen being alive, which makes an average increase of four each season. There is a further increase this season but to what extent I cannot say.

Arriving as they did in midwinter, these little animals, I can assure you, had a pretty hard time of it. However, after a few days' rest, having viewed the situation, they set vigorously to work to make themselves comfortable, and began to construct a dam by forming a dyke or embankment across a small moorland stream running through the enclosure; at the same time they commenced to build a house to live in.

The materials of which the dyke is constructed are wood, grass, mud and a few stones which are used for the purpose of keeping the grass and smaller pieces of wood in their place until more is built on top of them. They have continued raising this embankment to a certain extent every year, until it has now attained the following dimensions, viz.:– length, seventy feet; height in the deepest part, fully eight feet; breadth at base at deepest part, from fifteen to twenty feet, sloped inside, not straight across, but finely arched against the stream, so that it may the more easily resist the great pressure of water which it has to bear; – perfectly level, so that when a spate of water comes down it may run evenly over the top from side to side. So substantially have they built it, that no material damage has occurred to it from all the floods that have passed over it. They use a number of the larger pieces of wood as props, by fixing the thick end into the ground and the small end on the top, then build on the top of these, so as to fix them firmly. It would require to be seen to appreciate the great skill displayed in its construction; as I think it would tax the energies of a Bateman or a Gale* to make a better with the same materials. If any damage does occur, they immediately find it and repair it. I have seen them swim along the edge of the embankment, carefully examining it to ascertain the part most needful of

---

* John Bateman and James Morrison Gale, engineers who supervised the construction of a new water supply for Glasgow from the Trossachs lochs of Loch Katrine, Loch Venachar and Loch Arklet, including several dams.

repairs, then go to work with a will to rectify it. The dam is now seventy-eight yards long of still water.

Besides the dam already mentioned, upon which they bestow great care in its construction, owing to the house being built in it, they have other seven, some larger, some smaller; one of them having an embankment 105 feet long, and an average depth of three feet. These serve as places of refuge if the beavers are disturbed when out roaming about in quest of food or felling the trees, also as a waterway for conveying their food when storing it for winter.

In the construction of their dwelling the same kind of materials are used. As to how they built it: you must understand that for a considerable distance along one side of the stream, or burn, the ground rises in a steep bank, but about twenty yards above where they began to build the embankment for the dam there was a small level spot which they selected. Then at the bottom of the water they burrowed in three or four feet, rose up eight or ten inches, scooped out a space large enough to hold themselves, broke a hole in the surface about six inches in diameter, then began to cover it over with sticks, grass, and a few stones, always keeping it open in the centre by placing a few sticks perpendicularly, so as to act as a ventilator, and as the water rose in the dam and the family increased, they continued to build and enlarge the house, cutting their way up and forming their chamber or chambers inside, until it had now attained the following dimensions at the surface of the water (which is here about four feet deep), viz.:– height, about five feet; length and breadth about nine feet, having a door at both sides placed at the bottom of the water so as to prevent their natural enemies from following them, chief among which is the wolverine, although happily for both them and us there are none of these here to disturb them.

It is out of the water they take the materials with which they build their house. Were the sides of the house perpendicular they could not land; to obviate that difficulty they built a slip from two to three feet broad at its base, except where the doors are, so that they can land easily, and if they wish to enlarge the house they have got the foundation ready. To secure them against winter storms, they commence about the middle of September and give their house a coat of mud all over. It is with the mouth and forefeet, which are formed more like hands than feet, that they convey the materials of which their embankment and house are made. They do not use their tail, as was at one time said, for plastering on the mud, but their forefeet, with which they very carefully stow it in among the sticks. As to what they use for a bed to lie on, it is wood shavings, which they prepare in the following manner. After using the bark for food, they then place the stick on end, holding it both feet a bit apart, then with their teeth pare it down into fine shavings. They are very cleanly in their habits, as they often clean out their house, not casting away the refuse, but using it either on the top of the house or the embankment of the dam to patch up a hole.

Their food in winter consists wholly of the bark of trees; had they a choice I have no hesitation in saying they would prefer the willow and poplar. These not growing in the enclosure they had just to adapt themselves to circumstances, and take a share of what trees they could get, consisting of oak, plane tree, elm, thorn, hazel, Scotch fir, and larch. Of the hardwood, they seem to prefer elm to plane tree, then oak, of which they eat sparingly. Of the firs, the Scotch had the preference; as for the larch, they did not touch it till early in 1878, since which time they have taken to it very well. As for the alder and spruce fir, they eat almost nothing of them. Along with all these, we have always given them a supply of willow. In summer they eat freely of the common bracken, likewise grass, and young shoots of every description growing in the place. In autumn they grub up and feed upon roots, chief among which is the tormentil, and the young tender shoots of the common "spurts" before they appear above the ground, at the same time cutting down a tree now and again and feeding on the bark.

As to the tree-felling it is all done at night; the number of trees which they have cut down amounts to 187 trees from five feet in circumference downwards. These are all forest trees, besides a great many smaller bushes. Before cutting down a tree they mark it all round at the height at which they wish to cut it. They begin to cut at the opposite side to which they intend the tree shall fall, invariably making it fall with the top to the water. Where they grow near enough they make them fall across the stream or dam, causing many to suppose that they are so placed to form a bridge, whereby they may cross from one side of the water to another. They do not require a bridge, they can swim, and rather than cross over a prostrate tree they dive under it. My impression is that they are so placed to break the current of the water when the stream is flooded; also if convenient they take advantage of building a dam where some of the trees lie across the water. Those lying across in their principal dam are utilised in storing up their winter food, those stores being built on the upper side of the trees, so that they cannot be swept away with the winter flood.

When cutting the trees they use their teeth, on the same principle that a forester does an axe, always keeping plenty of open space, so that they can cut past the centre of the tree on one side before beginning on the other. It is in the latter end of autumn they commence to cut down trees for winter food.

Having cut them down, they speedily strip off the branches, cutting them into lengths to suit their strength for dragging them away to the dam, where they store them in different places near their house, so that they may have sufficient food, although the dam may be frozen over, or the ground covered with snow. What is left of the trunks of the trees that they cannot drag away, they feed on at leisure, eating the bark.

Besides the work above ground, they have done a great amount of underground work, such as cutting channels in their dams and making burrows. These burrows they make by cutting a road from the middle of the dam for several yards into the dry ground, where they scoop out a dome-shaped burrow from eight to ten inches above the level of the road, then cut a hole through the surface and cover it over with sticks and grass so as to act as a ventilator. Here they live and feed in security and contentment. Some of the roads to these burrows are from fifteen to twenty yards long and so level that the water follows them in the whole length.

As to the time they bring forth their young, from my knowledge I cannot say. I have seen it stated to be January, and also the beginning of May. I can say nothing against that, judging from the size of the young when I first saw them in the second week of June, the oldest litter being about the size of a full-grown rabbit, and the youngest not half that size.

From careful observation, I have good reason for believing they have only one at birth. One thing I am certain of, they have two litters in the season. Beavers are a class of animals that are very timid, their sight, scent and hearing very keen, so much so that it is with great caution they can be approached near enough to see what they are doing. They are under cover all day from seven o'clock in the morning till seven in the evening. When one comes out, it floats on the surface of the water, carefully surveying the whole scene around, sniffing the air, and if no danger is apprehended it dives and disappears. In two or three minutes, a number of the colony begin to appear and disperse themselves, some to swim and sport about in the dam, while others go in quest of food. If one of them espies danger it strikes one sharp, loud stroke on the water with its tail, when all of them that are out come tumbling into the dam and disappear.

I have seen them wrestle in playfulness and fight in anger, and also when the mother was feeding and the young one sporting about in the dam, I have seen it go and begin to tease her, when, if she did not wish to be troubled, she would strike and shake it and pitch it from her in the dam. They will allow of no laziness in any member of the colony; if any such there be, they are beaten and driven out to live as best they may. These so driven out generally roam about, making a burrow here and there, where they live for a few months and die.

Anyone who has had any dealings with beavers at all will recognise not just the essential authenticity of most of these observations, but also the skill and the dedication that has attended the work of writing them down, the innumerable hours of fieldwork over years that underpin these few hundred words. As well as appearing in the *Journal of Forestry*, Black's report appeared in at least two books, Harting's *British Animals Extinct Within Historic Times* of 1880 and Martin's *Castorologia* of 1892. The latter picks one or two holes in his observations, not least the final one about "lazy" solitary beavers, which were almost certainly sick rather than lazy.

But there is still no way of knowing whether the Marquess was just flaunting some of his wealth on assembling a meaningless collection of animals from the furthest flung corners of the empire (wallabies, for heaven's sake!), or whether, in the case of the beavers, he was actually interested in restoring the species, in which case he was a man some way ahead of his time. His choice of setting for the beavers' enclosure – "between Meikle Kilchattan and Drumreach" – is interesting, for it places the endeavour two miles south-west of Mount Stuart on a high, boggy moorland, rather than in the grounds of Mount Stuart, where they could have been shown off to visitors. So perhaps a very early conservation experiment was under way to try and establish whether reintroduced beavers could once again be a viable part of the Scottish landscape. If so, it failed utterly. It seemed not to have occurred to anyone to bring Eurasian beavers to Mount Stuart, given that these were the native species rather than the Canadian ones, but perhaps they were much harder to find, having become more or less extinct in western Europe. Or perhaps it was just that Mr Jamrach did not distinguish between the two and found Canadian beavers much easier to come by.

Castorologia records that a visitor to the enclosure in 1889 described it as "ransacked" and that "trees covered the ground in all directions giving the place an appearance of desolation". Reading between the lines it is clear that by then the beavers had eaten and felled themselves out of habitat, for such is the inevitable consequence of enclosed beavers with no management plan and no opportunity to move on to pastures new.

The project fades frustratingly from view in the Mount

Stuart archives, after Jamrach's letter of 1888 when the sea captain he had been relying on sailed off into the sunrise bound for Germany with his cargo of doubtless sea-sick beavers at £12 a head, a price the dealer declined to pay because it would have left him too little profit. By then, Joseph Stuart Black had died and the project seems to have bitten off more than it could chew.

I made the return ferry crossing and the drive home with a single train of thought for company – why beavers? The 1870s were around the high watermark of the Victorian killing years. Highland estates and their rich patrons routinely blazed away at birds with hooked beaks and mammals with teeth and claws. The white-tailed eagle, the osprey and the pine marten were already well on their way to extinction. The wolf was gone perhaps a hundred years before, or at least it was recalled only in the outermost edges of human memory. The beaver was gone probably 200 years before the wolf, having been no less systematically exterminated, albeit for different reasons. But although the wolf killed deer and was further burdened with 2,000 years of man-eating myth, while the beaver killed only trees and ate only bark and leaves (but in death made wonderful hats for style-conscious Europeans), both their exterminations were rooted in the same widespread contempt among the landowning classes of the day (up to and including the royal family) for anything at all in nature that could be killed for profit or sport or to protect their interests. Yet in the midst of such pervasive and grotesque hedonism, the Marquess of Bute was spending money and time and resources and manpower on a project that seems to have been motivated by – at the very least – a kind of admirable curiosity about beavers.

Then there is the nature of Black's report, which is essentially scientific. There is not a whiff of exploitation, no hint of a farming mentality with fur as an end product. Rather, there is an intimate portrait of beaver lifestyle within these admittedly constraining conditions, and an attempt to understand why they do what they do. And why produce a report at all unless the Marquess's motivation was to explore the possibilities of restoring the beaver as a native animal in the landscape? And surely (I'm guessing again) it would have been the Marquess himself rather than his keeper who initiated the report and arranged for its publication? Alas, not even the National Library of Scotland could shed any light on the *Journal of Forestry* in the 19th century, so all that survives is that lengthy extract, with no explanation of the thinking that underpinned the project.

So why beavers?

Why isolate them two miles away?

And why the report?

I think now that I may know why. The most rudimentary enquiries into the biography of the 3rd Marquess of Bute reveal nothing at all to suggest an inclination towards pioneering nature conservation. Instead, it is blindingly obvious that the passion that outweighed all his other far-flung fields of interest and expertise was – architecture!

He used his colossal family wealth, mined in the coal-fields of Wales, to embark on a series of extraordinary architectural ventures. In particular he formed a partnership with the London-based architect and designer and champion of the Gothic revivalist movement, William Burges. Together they insinuated spectacular Victorian mansions onto the heroic landmarks of first Cardiff Castle (an 11th-century castle the Bute family had acquired in the mid-18th century), then nearby Castell Coch, a 13th-century ruin. And Burges, as much as anyone, pre-figured the Arts and Crafts movement, which, in turn, found one of its most original flowerings in Charles Rennie Mackintosh and the Glasgow School.

After the 1877 fire at Mount Stuart, the Marquess's admiration for the work of Burges must have influenced how he envisaged a restored house, and directed him towards Robert Rowand Anderson, who was among Scotland's pre-eminent architects of the day. Anderson built something remarkable at Mount Stuart, and you can see how it influenced his Scottish National Portrait Gallery in Edinburgh a few years later.

Architecture, then, was the great love of the Marquess's life, and can it be coincidence that in the midst of the two great projects with Burges, he brought nature's architect – the beaver – to Mount Stuart? Look at Black's report again. I fancy the Marquess had a hand in it; I can see him, pen in hand, reading through Black's draft, and inserting a pointed aside that interrupts the otherwise comfortable flow of Black's narrative, at the point where it discusses the beavers' embankment: *It would require to be seen to appreciate the great skill displayed in its construction; as I think it would tax the energies of a Bateman or a Gale to make a better with the same materials.*

You might not expect an estate keeper on Bute to be able to name-drop the dam-builders of the Trossachs lochs, but the far-travelled Marquess with a passion for architecture probably knew them by their first names.

So why beavers? Architecture. That's why.

Chapter 5

CASTOR FIBER:
AN UNNATURAL HISTORY

THE EURASIAN BEAVER (*Castor fiber*) is Europe's largest rodent. The Canadian beaver (*Castor canadensis*), its transatlantic kin and to all intents and purposes the same animal, is North America's largest rodent. These may not be the kind of slogans that soothe public perceptions and win elections because many people associate "rodent" with "rat" and a tree-felling rat is surely a nightmarish beast such as Julia Donaldson might have conceived but ditched because it was too scary for *The Gruffalo*.

But the beaver is no rat. This is the animal that gives rodents a good name. One of its most deserving and eloquent tributes was paid by the American writer Ernest Thompson Seton in *Wild Animals at Home* (Doubleday, New York, 1913):

... the Beaver is the animal that most manifests intelligence by its works, forestalls man in much of his best construction, and amazes us by the well-considered labour of its hands.

There was a time when the Beaver's works and wisdom were so new and astounding that super-human intelligence was ascribed to this fur-clad engineer. Then the scoffers came and reduced him to the low level of his near kin, and explained the accounts of his works as mere fairy tales. Now we have got back to the middle of the road. We find him a creature of intelligence far above that of his near kinsmen, and endowed with some extraordinary instincts that guide him in making dams, houses, etc, that are unparalleled in the animal world...

The chances are you have never seen a beaver because most people have not and most people won't, given its dusk-till-dawn night shift predilection. So if you have not seen one, the beaver's vital statistics are these: body length, 75–90 cm; tail length, 28–38cm; weight 15–38kg. If that doesn't mean too much to you, at its chunkiest its nose-to-tail length is a yard-and-a-half, and it possesses the kind of power-to-weight ratio that can fell, say, a forty-foot tall tree with a trunk fifteen inches in diameter, and, having felled it, lopped its branches, stored (or eaten) its foliage, manipulated the trunk (whether intact or reduced to bite-sized chunks) improbable distances through still or turbulent water to its final destination in whatever architectural project it is pursuing at the time. On the other hand, it may just leave the trunk where it lies and return for it much later. Or it may forget about it altogether. This is because not every tree the beaver lays siege to is necessary for beaver architecture. Sometimes the beaver is just servicing his tools, specifically his incisors. These simply never stop growing, so they must be worn down at the same rate at which they grow, and the way they do that is to bite wood, much more wood than they can possibly use as building materials.

So every beaver landscape has its felled but discarded trees, often lying half in and half out of the water as though indecision about its fate had so tormented the architect that the problem of what to do next had proved insurmountable. And every beaver landscape has its standing trees which have been eaten part of the way through the trunk and then abandoned, as though the beaver had changed its mind about the trees in question after several hours or even days of biting and reducing two large wedges on opposing sides of the trunk to piles of vivid white chips of wood which lie tidily at the base of the tree. Confronted with such a landscape, human observers are apt to throw up their hands and wail about wanton destruction, about vandalism, about loss of native trees for no good reason, and in the throes of such wailing, to harden their hearts against what they consider to be this most unwelcome of intruders.

The trouble with that interpretation of events is that it is based not on any knowledge of how the beaver works but rather on the basis of 400 years of human history in a landscape devoid of beavers. All our understanding of the way river systems work is based on river systems without beavers. River systems with beavers, which are nature's preferred option across the entire northern hemisphere, dance to a very different tune. So today, any landscape anywhere in Britain that has recently acquired beavers also has 400 years of catching up to do. It is no surprise then that many of the landscapes beavers have been able to choose for themselves in the last few years (as opposed to the official trial reintroduction in a landscape chosen by people) are historic beaver landscapes too. Even after 400 years, landscapes once suitable for beavers don't stop being suitable. In fact, especially after 400 years, for one of the things to bear in mind is that beavers work on a different timescale to people; they take a longer view of things and every beaver generation contributes to a cycle that goes on repeating down the centuries, a slow rhythm and almost imperceptible to human sensibilities (unless the humans happen to be archaeologists perhaps), a rhythm that permits constant renewal of the beaver's world.

This is how it works. Despite the fact that beavers like to surround themselves with comparatively fast-growing trees like willow, birch, alder and aspen whenever possible, as often as not they will eventually eradicate their own supply of food and building materials. If this happens in the wild, the beavers simply move on. If you were to chance on that landscape the day after they had departed, you might feel justified in declaiming such desecration, but nature and the beavers know what they are doing, and you may not because you have not seen what comes next. First, the beavers' dams begin to collapse. This often happens very slowly, because of the quality of workmanship that underpins the architecture. Over months, years and decades, the canals and pools and ponds spread and degenerate into that most precarious and precious state of landscape we call wetland. By then, many of the leftover tree stumps have begun to resuscitate, to regenerate into low scrub. Surrounding woodland begins to advance on all sides, new generations of the very water's-edge tree species the beavers had targeted. And in time, in time... because the beavers did their work well, and the land-and-waterscape has prospered in their absence to become suitable for beaver habitation once again... in time and countless generations later, the beavers will return. Once again this very landscape will accommodate new dams and lodges, and this time there will be more open water for the beavers to work with. This is the benevolent rhythm to which beaver life moves, a constant cyclical process of expansion and renewal, and with each stage of the cycle, the beavers create new opportunities for plants, insects, birds, fish and other aquatic life, and water-loving mammals.

That reference in *Castorologia* to the Bute beavers enclosure looking "ransacked" and offering "an appearance of desolation" can be explained because the beavers were walled in, their numbers increased steadily, there was no management of the numbers, and there was no opportunity for the beavers to move on once they had consumed all the natural resources available to them. So the worst time to try and assess the impact of beavers on the landscape is when they have just moved out. Come back in twenty or fifty years, examine their true legacy, and then decide. But the trouble with our relationship with nature conservation is that we don't have that kind of time, or at least we don't often think in that kind of timescale. We want immediate results and in circumstances where nature is not immediately forthcoming we curse nature's cussedness and look for scapegoats to control or cull or render extinct. It's what our fathers did and our fathers' fathers and theirs and so on; and those generations that did know about co-existence with the other creatures of the earth (often holding some as sacred, some as teachers, some as both), are so distant now as to be far out of reach to our own 21st-century sensibilities.

The phenomenon is nowhere more evident than in the angling lobby, which has been loudly deriding the spread of beavers across the Tay catchment as a dam too far, and arguing that permitting them to carry on living there is a risk too toxic to be worth taking, a disaster-in-waiting for the phenomenally lucrative salmon fishing season, all this without being willing to watch what happens for a few years to see how their fears measure up to the reality, without being willing to acknowledge the shortcomings of rivers that have evolved for 400 years without beavers.

A place where they know a thing or two about wild beavers and wild salmon and the subtleties of how they co-exist is British Columbia. A four-page spread in *British Columbia Magazine*'s issue for winter, 2013, entitled "Leave It to Beavers", listed "ten reasons to praise B.C.'s eco-engineers". Number six on the list was, "They protect salmon fry." The article went on:

Fisheries managers used to destroy beaver dams because they thought they interfered with salmon migration. Now they're learning that beavers actually improve life for salmon and other fish. Beaver ponds provide excellent rearing habitat for juvenile salmon, with ample food and good cover to protect them from predators.

Beaver dams generally don't restrict fish movement. Adult salmon can leap over most dams, while smaller fish just wriggle through, because a beaver dam is not like a human dam. It's porous. Water flows in and out. There are places for a fish to go through.

Oddly, this put Glen Orchy into mind. Glen Orchy in Argyll is one of my favourite places to watch salmon on the move. The River Orchy's waterfalls are not show-stopping

cataracts, but rather they are long and complex and comparatively shallow. Salmon have no trouble with the height
of the obstacles, but their first leap inevitably lands in the
midst of the chaotic tumbledown of huge rocks and small
mini-falls within falls, chutes and pools and whirlpools, and
the thunderous press of the river. There are difficult ways
up the falls that the strongest fish can accomplish in two
or three leaps, and there are wriggle-routes for the small-
fry. I never thought about it as having the properties of a
beaver dam before, but now that I have become slightly
conversant with a range of beaver dams, beaver canals, and
beaver pools, I can see only benefits for salmon and trout,
and therefore also for the angling fraternity. But it must first
be willing to examine the slow-burning effects of beavers,
now that they are back.

If you read a book like *Decade of the Wolf* (The Lyons
Press, 2006), Doug Smith and Gary Ferguson's account of
the first ten years of wolf reintroduction into Yellowstone
National Park, the impact of the wolves' presence on the
ground was almost immediate and far-reaching, and produced the kind of headlines that conservation dreams are
made of. They set in motion a phenomenon called a "tro-
phic cascade", by which the presence of one species at the
top of the food chain generates a benevolent chain reaction
that travels down the entire length of the food chain, an
event that begins to be measurable and clearly visible within
a handful of years. That doesn't happen so obviously when
you are as far down the food chain as beavers. What you
see instead is a lot of felled, half-felled, broken trees, an
"untidy" half-sunk landscape that has begun to unpick the
seams of the order the beavers found when they moved in,

the order that was achieved by 400 years of rivers without beavers. What you don't see, unless you are willing to sit still in the midst of their landscape and look closely, is that new opportunities have already begun for some of the other creatures that live there, and some that didn't because until the beavers came along they had no foothold to work with.

The truth is that beavers and wolves have a lot in common. In a new landscape, or rather a very old landscape they have reclaimed having once been historically cleared from it, they begin to reshape what they find. The wolf's immediate impact is achieved by its dramatic effect on browsing herds so that these are constantly on the move instead of sedentary, with the result that browsing a landscape to the bone is no longer possible and regeneration rushes in, regeneration that galvanises and magnetises nature's waiting hordes. The beaver achieves transformation by its capacity to generate wetland then expand it, and to clean up the rivers it colonises because its architecture (particularly its dams) acts as a series of underwater sumps to which bacteria cling and are slowly broken down, a phenomenon which is already measurable in Scotland, although it is not necessarily the kind of phenomenon that makes headlines. Number seven in the *British Columbia Magazine* article was, "They purify streams":

Wetlands have been called the "kidneys of the landscape" because they work to purify the water that passes through them. Beavers help to keep these vital organs functioning. When a beaver pond slows the flow of a stream, sediments and pollutants settle out, cleansing the water before it continues downstream. Meanwhile, pond plants and micro-organisms break down potentially harmful substances that are left behind.

Beaver and wolf are innovators for nature, and it is no coincidence that one of the species to benefit most and almost immediately from the wolf's return to Yellowstone was the beaver. There had been only one beaver colony inside the national park before the wolves came back. But once regeneration of the fastest growing trees (willow, birch, aspen, cottonwood) got going, the beavers from just outside Yellowstone were among the first to notice the change. They began quietly following the wolves wherever they went, scattering wetland as readily as the wolves scattered the deer herds. And following the beavers came water-loving plants, insects, birds that prey on the insects, and birds that like feeding on underwater plants like swans, geese, and a myriad other life-forms including frogs and otters, turtles and moose. Within seven years there were nine beaver colonies.

And here's another thing. Apart from human beings, guess which creature is the only one in all nature that purposefully traps and harnesses water for its own use. So when I write that the beaver is Europe's biggest rodent, that's a good thing, right?

☉☉☉

The nadir of beaver fortunes across much of the northern hemisphere was around the beginning of the 20th century. The scale of the slaughter up until then had been breathtaking, first in Europe and more sporadically through Asia, and then (once European colonists began importing their casual ways with nature) throughout North America. No-one knows what the populations of the two beaver

species were before the slaughter began, but current esti-
mates for North America before the Europeans arrived
are between 60–400million. By 1900 that number had
fallen to about 100,000, and in Europe and Asia a population that must once have also numbered many millions
was reduced to around 1,200 animals in a handful of far-
flung and isolated locations. There is no single explanation for why the rot stopped, but one factor was surely
that "harvesting" beavers in ever smaller, more isolated
and more remote colonies was no longer worth the effort.
The slaughter stopped because it was suddenly too much
like hard work for too little reward. The very first attempt
at turning the tide was made in Norway as early as 1845
when hunting beavers was banned. From what we know
now, thanks to conservation programmes across the northern hemisphere, beaver populations respond quickly to a
second chance. The scale of habitat destruction and the
expansion of human populations have already guaranteed
that those once huge beaver populations are a thing of the
past, but in North America the Canadian beaver is advancing strongly again, and in a well-studied area like British
Columbia, for instance, it is back up to around 500,000.
The Eurasian beaver is also prospering. In 2014 the IUCN
Red List of Threatened Species showed a dramatic rise
throughout its range (from western Europe to China) from
430,000 in 1998 to 639,000 in 2006, a figure it described
as "almost certainly a considerable underestimate". In
France, numbers have risen from near-extinction to around
15,000, in Germany from near-extinction to around
30,000. Altogether, there have been reintroductions in
(deep breath) Austria, Belgium, Croatia, Czech Republic,

Denmark, Estonia, Finland, France, Germany, Hungary, Italy, Latvia, Lichtenstein, Lithuania, Montenegro, the Netherlands, Poland, Romania, Serbia, Slovakia, Slovenia, Spain, Sweden, Switzerland and Ukraine.

Given the pool of knowledge that exists in the wake of such widespread conservation measures, it makes Scotland's official five-year trial in Argyll with a handful of beavers look rather timid. As it is, the un-tagged, un-chipped, free-range, and diligently exploratory Tayside population is now so well established and so demonstrably thriving that it would be good to think that the point of no return is finally far behind us and retreating ever further far back under the horizon. It certainly will be by the time the Tayside beavers arrive in Argyll and start mingling with the beavers from the official trial.

Meanwhile, in England and Wales, there have been other beaver stirrings. Through the first decade of the 21st century several nature reserves and county wildlife trusts opted to recruit beavers in controlled conditions to foster wetland and increase biodiversity. These included Ham Fen in Kent, Lower Mill Estate near South Cerney in Gloucestershire, Blaeneinion in Wales, the Wildfowl and Wetland Trust reserve at Martin Mere. Then, early in 2014, beavers were found to be breeding on the River Otter in Devon, a discovery that produced a typical response from the Department for Environment, Food and Rural Affairs (Defra). The beavers would be captured and removed to avoid the risk to human health from a tapeworm that beavers may carry. Given that this came hard on the heels of the notoriously ill-judged bovine T.B. badger cull in the south-west of England, there was more widespread outrage.

Happily, sage counsel prevailed. Devon Wildlife Trust, aided by overwhelming community support, spearheaded a pro-beaver campaign, Defra's plan was abandoned, and following a year of campaigning, monitoring and thoughtful management, English Nature granted the trust a five-year monitoring licence. In 2020 the River Otter Beaver Project will present its findings to Natural England for a final decision. By then, it would be nice to think that the Scottish experience had demonstrated the overwhelming array of natural positives that flow where wild beavers prosper.

All of which poses the question: why the relentless, globe-encircling bloodlust for beavers in the first place and for so many hundreds of years? There are two answers, the first of which is fur. Beaver fur is luxurious and double-layered. From the robes of kings to the hats of Europe's thirst for high fashion, it was in demand for a thousand years, and by the time Scotland finally ran out of the very last British beavers, probably in the 16th century, the unsatisfied demand simply looked elsewhere. It looked in particular to Canada. *Castorologia* intoned:

> The great slaughter began with the establishment of the first fur trading post in 1604, when Champlain planted his colonists at Quebec, and followed with other settlements on the St Lawrence, which from subsequent experience proves to have been the natural highway to the richest fields on the continent…
>
> The Dutch from New Amsterdam and the neighbourhood of the Hudson River, traded also into the lake district and helped materially to thin the numbers of the beavers, from which followed contention and conflicts with those who tried to control the Indian trade in the rich peltries.

The second answer to the question why is oil. The beaver has two scent glands that secrete a chemical called castoreum, and the discovery that it had medical properties only accelerated the rush towards extinction. I should confess at this point to a degree of culpability, or at least my mother inflicted culpability upon me, for one of the worst fates that could befall me as a child was to be confronted during some minor illness or other with a bottle of castor oil. Regardless of how sick I might have felt, nothing could induce me to throw up more enthusiastically than being compelled to swallow a spoonful of castor oil, which is the foulest-tasting medical concoction on earth. Beavers of the world, all I can say in my defence is that I never knew, and I'm pretty sure my mother never knew either. Quite apart from the trivial reasons for which it was mercilessly prescribed to me, castoreum has pain-killing properties, is an anti-inflammatory, and can help to raise blood pressure and cardiac output. Other reputed uses for castoreum down the years (and hopefully at least some of these are surely urban myths) included flavouring cigarettes and persuading bees to produce more honey. Ye Gods! Urban mythology is silent on how that particular persuasion was effected.

The inevitable expansion of the white man's regime in North America doomed ever more beavers, echoing precisely what had already been achieved in much of Europe with the Eurasian beaver, and for that matter echoing precisely what had already been achieved in much of Europe with the wolf. The fates of both animals were on parallel paths. By the time *Castorologia* was published in 1892 the Canadian beaver must have looked every bit as doomed as its European kin, an expectation that is confirmed by the following bleak assessment of how things stood at the time:

The question is often asked, "Where today are beavers to be found in their primitive state?" and the answer is not difficult to give, for the beaver is of slow locomotion on the land, and its habits confine it very closely to the neighbourhood of its birth; it keeps to the watercourses, and as the hunters follow, it recedes farther up the streams, till on the height of the land, the quiet lakes and pools offer a last retreat, but alas no sanctuary: and the white man with his "fire waggon" dashes through the woods, changing as if by magic the country through which he passes, with utter disregard for the quiet denizens of the forest.

Then the drastic final warning:

As to the ultimate destruction of the beaver, no possible question can exist, and the evidence of approaching extermination can be seen only too plainly in the miles of territory exhibiting the decayed stump, the broken dam and deserted lodge. The passing bear or wolverine tears open the lodge, partly in the vain hope of finding a meal, partly from habit; the rising waters float the logs away, while the drifting ice in fall and spring gradually destroys the dam till within a decade, where once the busy colony spent their happy domestic lives, no sign remains of their wondrous toil.

But the beaver's headlong descent down the way of the dodo pulled up at the brink. And given that corrective action had already begun in Norway forty-five years earlier, it is just possible that the first hazy stirring of conservation thinking was also exported from Europe and found a toehold in the New World. On both sides of the Atlantic the seed was planted, but its fruits have proved to be a particularly slow-growing strain, and nowhere was more reluctant to embrace it than Britain.

By the time anything like reliable written records began to appear, the post-Ice-Age beaver had been in place for around 9,000 years, edging north as the great ice retreated from the face of the land. Evidence of their historical presence appeared rather miraculously in 2007 deep beneath the surface of Loch Tay in Highland Perthshire. Members of the Scottish Trust for Underwater Archaeology, based at the wonderful Crannog Centre near Kenmore at the eastern end of the loch, had discovered a drowned forest in Loch Tay two years earlier, but as exploration of the site progressed they began to find the remains of beaver dams and lodges. Carbon dating has since shown that the remains are between 1,500 and 8,000 years old. Individual pieces of wood show clearly the perfectly preserved toothmarks that are the unique signature of all beaver-felled timber and beaver-gnawed branches.

So, whether or not you are persuaded that beavers are rampaging vandals rather than agents for the long-term conservation of native woodland, consider this: the greatest extent of natural woodland cover in Scotland (often misleadingly referred to as "the Great Wood of Caledon") was around 5,000 years ago, by which time there had already been an unhampered and therefore burgeoning beaver population for 5,000 years, and these had rampaged to such a pitch of vandalism that the native woodland of the country had never been in better heart, and it never has been since. The true rampaging vandals that accounted for the demise of the trees were the same vandals who accounted for the demise of the beaver, the lynx, the brown bear, the wolf, and any day now must be held responsible for the more or less inevitable extinction of the Scottish wildcat.

Those reliable written records I mentioned above probably began with Welsh laws enacted around 940 AD, and which included several provisions concerning the relationship between people and wildlife. One of the most telling is related by Harting in his *British Mammals Extinct Within Historic Times* and using language that sounds as if it might have been written down in 940 AD too, instead of 1880:

> *It is there laid down that the king is to have the worth of Beavers, Martens and Ermines, in whatsoever spot they shall be killed, because from them the borders of the king's garments are made. The price of a Beaver's skin... at that time was fixed at 120 pence, while the skin of a Marten was only 24 pence, and that of a Wolf, Fox and Otter 8 pence. This shows that even at that period the Beaver was a rare animal in Wales.*

I think it shows no such thing, but rather that the beaver wore the best fur. If it seems an excessive demand to make, that all beavers "in whatsoever spot they shall be killed" were needed to make the borders of the king's robes, it should be borne in mind that making garments from animal furs is a pursuit that involves a high percentage of waste. I had a glimpse of the phenomenon at work in Alaska when I met a woman there who was preserving the age-old tradition of making ceremonial capes for traditional tribal dances, and teaching the skills to schoolchildren. She was using squirrel fur. She showed me a sample of a finished garment, which, she said, was made from the fur of forty-eight squirrels.

"You mean for every ceremonial cape, forty-eight squirrels have to die?" I asked.

"Oh no," she said, "sixty-four squirrels have to die. We have to make sure all the squares match up."

So I would imagine that if that is an issue among Alaskan tribes like the Tlingit, to which she belonged, then it was rather more of an issue for the King of Wales. Bear in mind too, that we have no idea how many beavers he was talking about, and that he would have had rather more than one set of robes, what with being king and all.

The question of when the beaver became extinct in Britain has never been resolved with any degree of satisfaction, but there seems to be academic consensus at least that it has been gone for about 400 years. The Scottish mediaeval historian Hector Boece insisted they were present around Loch Ness in 1526. But by the time Sir Robert Sibbald published his natural history work *Scotia Illustrata* in 1684 there was no mention of beavers. Sibbald was a serious establishment figure, a surgeon and antiquary and the Geographer Royal to King James VII. His fascination for geography and natural history resulted in a request from the king to write a natural history book about Scotland. His method was to send a questionnaire, complete with instructions about how to fill it in, round the great and the good of society. Clergy and Gentry were two of his selected categories in different parts of the country. His book was compiled from sixty-five respondents, none of which could report the presence of beaver. But that is hardly the same thing as saying that the beaver was extinct.

Like the wolf, the beaver would have retreated further and further from established human presence as its numbers dwindled down to the last few, and for that reason as much as any other, it is at least likely that they lingered on here and there in the quiet places of the land for some years and possibly some decades after the last time any human eyes could report a sighting.

Archaeology has begun to fill in some of the blanks on the historical beaver map of Britain, a geographical spread from Sutherland and Caithness in the north to East Anglia and Cambridge in the south (the Fens seem to have been something of a stronghold, as do the Scottish Borders). But one find particularly intrigued me – the bones of two beavers emerged during drainage operations in 1821 at Marlee Loch, also known today as Drumellie Loch, near the east Perthshire village of Kinloch a little to the west of Blairgowrie. The remains were found between five and six feet down in a peat bog, and between the layers of peat and marl. One of the finds was an intact skeleton, but it appears to have shattered during excavation. The skull survived, however, and is in Perth Museum and Art Gallery.

There is a rather grim irony at work here, don't you think? The beaver, the only manipulator of water in the natural world, the preacher of wetland gospels, was finally flushed from its last resting place as a skeleton, and by a drainage project of all things.

Marlee Loch lies deep in the heartland of the 21st-century Tayside beavers' expanding realm. For some reason that I imagine would surely attract the scorn of the partners of the official trial in Argyll (given that they have disapproved so heartily of the Tayside incursion), I found myself embracing a perverse species of satisfaction from the possibility that beavers had found their own way back to a site where it is known that they lived a thousand years ago, perhaps much longer. So I went for a look.

I found the place one day in a midsummer midday doze. An osprey hung on still air high over the middle of the loch, which I chose to interpret as a good omen, for it is another of nature's vanquished tribes that has lived to tell the tale and returned to haunt its old quarters. (My osprey associations stretch back forty years now: I was involved from a very early stage in watching a pioneering pair that settled at the Lake of Menteith in Stirlingshire in the 1970s. They were not only the first birds to nest anywhere in Scotland outwith Strathspey, but they were also the first birds to return to what had been one of their historic heartlands. This willingness in nature to re-colonise old embattled homelands, decades or even centuries after it has been crudely evicted, is what I like to think of as a kind of wild forgiveness. Nature's faith in its old homelands and our willingness to accommodate it now has been justified again and again, and now on any one summer afternoon you can watch a dozen ospreys fishing over the Lake of Menteith.)

A sandpiper piped, zipped out from a bare yard of shoreline in a wide, chittering arc that covered fifty yards but never rose so much as a foot above the surface of the loch, and joined its mate on a small rock in the shallows where the chittering stopped as abruptly as it had begun and they bobbed a greeting that had a curiously, formally Japanese-looking aspect to it. Sandpipers are a living, breathing, beautiful, bowing fragment of almost all my favourite shores. I was feeling good about this place even before the osprey, even before I focussed binoculars on the birds' rock. And of course, for such is often the way of these things, they vacated the rock in the same instant. And, for such is also the way of these things, nine times out of ten I would have panned the glasses across the loch to follow their flight, but this proved to be the other one out of ten, because some-thing in the glasses and some distance beyond the rock (so out of focus) had caught my eye, and instead of following the sandpipers I re-focussed the glasses on the base of a tree, a tree from which a large swathe of bark had been stripped. From the bright white colour of the wood (which is what had held my attention in the first place), the bark-strippers had clearly been at work very recently. I had come to the loch willing to spend as long as it might take – all day and all night if necessary – to scour the loch's surface, its shores and margins and nearby woods and burns, in search of evi-dence to try and justify my only slightly fanciful notion that beavers had returned unbidden to this small and insignif-icant loch, insignificant that is apart from its place in the archaeological history of beavers. As it happened, it took me a quarter of an hour.

Shortly after my visit to Marlee Loch, I learned that bea-vers had also found their way back to Loch Tay, where one day soon they may dive down into its depths and find the toothmarked signatures of their 8,000-year-old ances-tors. I find the idea immensely comforting and a cause for optimism.

Marlee Loch is one sun-smitten jewel on a necklace of small lochs that leads west from the confluence of the Ericht and the Isla, two of the rivers most associated with the spread of the Tayside beavers. I spent the rest of that day going from loch to loch all the way west to the Scottish Wildlife Trust reserve and celebrated osprey landmark of

Loch of the Lowes near Dunkeld. There the remarkably long-lived and profusely fertile resident female (whose many fans call her Lady) was so famous she had her own book and her own website, and her every move travelled the world via webcam and internet. She failed to return from migration in 2015 and was replaced at the nest by a new female, which was immediately christened Lassie. And while I think that this kind of reserve management sails very close to the wind of exploiting nature, there is a counter-argument that says it has won many friends for nature in general and ospreys in particular. But inflicting celebrity on wildlife rarely improves understanding, and I don't think it has done here.

I mention it because since 2012, beavers have also been swimming into the gaze of the Loch of the Lowes cameras, and have quickly become the focus of frivolous blogging. The day they get names and a website too, no-one should be surprised if there is an outbreak of mysterious signs at the neck of land between the loch and Butterstone Loch immediately to the east, mysterious because they will be written in beaver language and urging all west-making beavers not to cross into Loch of the Lowes but to take the high road north by way of the western slopes of the Knock of Findowie to the blessedly camera-free trinity of Mill Dam, Rotmell Loch and Dowally Loch and thence to the Tay, from where the wild world will be their oyster once again. Or, they could take the really high road by way of Benachally and Loch Ordie, and still join the Tay at Dowally.

There is a serious point to this frivolity. SWT is one of the partners in the official trial in Argyll, as is the Royal Zoological Society of Scotland, and you may well wonder why a wildlife organisation is hand-in-glove with an organisation that runs two zoos – the Highland Wildlife Park at Kincraig on Speyside and Edinburgh Zoo, where the society's giant panda enterprise is just one of many reasons why I detest zoos and everything they stand for. As an organisation, SWT has championed the official trial in Argyll as the only appropriate way to countenance a reintroduction, and has disparaged the unofficial and infinitely more successful reintroduction on Tayside. Yet, oblivious to this state of affairs, the Tayside beavers have washed up in the waters beneath what most people still think of as Lady's Tree, despite the fact that Lady has been usurped by Lassie. It is as Lady's Tree that it has become a cross between a shrine and a totem pole among the more populist extremities of nature conservation in Scotland. And there, nature's architect, and peerless tree-feller, has begun to breed, and for that matter has begun scene-stealing on the webcam. (I haven't checked, but I am sure it must have occurred to someone in the SWT to wrap the lower reaches of Lady's Tree with several layers of the toughest protective coating that mankind can devise, just in case the beavers take a notion to sink their teeth into the very tree that was named as Scotland's inaugural Tree of the Year in 2014.)

My nightmare is that common sense fails to prevail in the matter of national decision-making about the future of wild beavers, and the small-mindedness and self-interest of the landowning and angling lobbies prevail instead, and Edinburgh Zoo and the Highland Wildlife Park suddenly acquire a healthy stock of beavers, and a sunlit shaft of enlightenment that blazed briefly from coast to coast is extinguished.

Chapter 6

THE RIVER DIARIES: II

BEAVERS BY MOONLIGHT seemed like a good idea. A string of clear nights either side of a full moon offered me a gilt-edged invitation. Moonlight gilding every corrugation in the river, gilding every ripple on pool and canal, layering wet flanks of beaver fur with a gilded dazzle, every newly-gilt detail of their architecture aglow even as they muddied it into place, gilt-edged cubs playing with their own moonshadows, new avalanches of woodchips at the base of a gnawed tree rendered into white-gold heaps of unburied treasure. That was the kind of thing I had in mind, the transformative effect of moonlight on the often unrewarding evening shifts of occasional super-wary dusk appearances, then much peering at the tree-shrouded black river for black beaver shapes practising the black arts of their architecture in the dark. I made the accustomed drive across the moor towards my chosen Beaverland humming *Moon River*, and dreaming Audrey-Hepburn-shaped daydreams.

In the early days of my old badger-watching exploits, all manner of badger-watching sages and veterans had emerged from the woodwork, determined to help, brandishing mantras. One was that badgers *never* come out in moonlight because it makes them feel insecure. I used to listen to these people more than perhaps I should have, but after the first time I saw a badger out prowling under a full moon in an open field, I must have encountered them in a dozen different shades of moonlight and phases of the moon. Since then, anytime I come across a pet theory about wildlife that includes the words "never" or "always", I treat them with deep scepticism. The other mantra – the "always" one, as opposed to the "never" one – was: "Always get there an hour before sunset." I had put that one into practice for weeks before it occurred to me to pose the only relevant question: why? Eventually, I found my own ways of working badger setts and badger territories and none of them involved getting there an hour before sunset, and I moved to my own utterly trustworthy mantra: "Get lucky."

I was thinking about that in particular, and similarities between watching badgers and watching beavers in general, because, as it happened on the evening of the beavers-by-moonlight expedition, I got there an hour before sunset. I had not planned it that way, it was just how the day panned out. As with the badgers, it did me no good, and as with the badgers, I got lucky.

I watched the western sky fire up then pale as the land darkened, watched the purples fade to blue-blacks. The woodland darkened, the banks and the canal and the pools and the dams and the river all darkened, and I sat and watched the darkening and waited for the moon. The evening was windless and quiet apart from a turbulence of cows, a stupidity of pheasants (moments of idleness in nature's company often generate new and imaginative collective nouns), and the many-voiced haverings of the river. There are worse conversationalists for a nature writer with time on his hands than a Highland river. But still the land only grew darker.

It took longer than it should have for me to appreciate that only the fields to the west, glimpsed through trees, were in moonlight. The same dullness of thought finally realised that the thickly wooded hill at my back was an effective barrier to moonlight and a source of colossal shadow. It was at that point that I decided to back an outrageous hunch. It was unsupported by anything like sound reasoning, and biology certainly had nothing to do with it, but rather it clung tenuously to the fact that once, twenty years before, I had disproved the folly of the badger sages in the matter of moonlight and other clichés. I guessed – wildly – that if badgers were not averse to foraging in moonlight, there was no obvious reason why a beaver should not be either. And if the particular nature of this particular beaver territory pre-cluded the possibility of moonlight even on a moonlit night, perhaps a beaver or two might go curiously out into the open moonlit country, rather in the spirit of Mohammed going to the Mountain, the better to see what was going on in the neighbourhood and why.

As I have already said, I am a nature *writer* rather than a naturalist, and I confront dilemmas in nature's company with a writer's instinct rather than that of a naturalist, for that way (as I see it) lies the path to a better book. If you were to ask how that instinct might manifest itself, this would be the kind of thing I have in mind.

So I walked away from the shadow-shrouded heart of Beaverland just as I had done the day I saw the kite squad-rons, out into the first of the open fields where the moonlight

flooded the land with the enthusiasm of dam-building beavers, and my moonshadow leapt into life. I walked down
the field-edge towards the river that had loped away north
in a long curve just west of the trees, and was suddenly
distant. A shrub-darkened burn oozed down the side of the
field, and I reasoned that although I was as moonlit as anything else out there, my shape would be against the darkness
of the shrubs and clusters of small trees, and that would
perhaps diminish my impact on the landscape in the eyes of
the watching night tribes, and the mutter of the burn might
deaden my footfall.

There are trees along the top of the riverbank here, but
they are few and far between and their crowns are blasted by
winds and their roots undermined and tugged at by spates.
There is not much to sustain beavers here, although half a
mile further west the trees gather around the river again and
there is new beaver life in those trees too. As I walked softly
along the bank, I noted the occasional patch of stripped bark
low on a trunk, and one thin branch lopped off in telltale
pencil-point fashion, and these suggested that beavers passed
this way and occasionally paused but did not linger.

The river was quieter here, except where it parted
with itself to surround a bank of shingle in midstream and
fussed and bounced in dozens of tiny waves as its two limbs
reunited at the shingle bank's downstream end, and every
one of those tiny wavetops wore a dancing crest of white
gold. Even if nothing else happened, it was a wondrous
night to be out.

I considered the shingle bank from some distance away,
simply because it was a different landform, something different to look at, that and the fact that a similar bank a

couple of miles downstream was decorated by a discarded slim branch that a beaver had obviously been working on. And from what I heard elsewhere, beavers are drawn to any kind of natural "island" in their home waters.

Just ahead of me there was a single hawthorn bush in its scrawniest midwinter guise. It was about my own height, and although it was as transparent as cellophane it was handily placed and might just serve to break up my silhouette. I approached it, then, with an already stirring mood of expectancy. Then something out on the shingle moved a few inches, then slid back into position, an awkward enough movement to command my attention and rouse my well-submerged naturalist instinct. A few days before, I had bought a new pair of binoculars and been persuaded to pay a bit more than I had planned because of the particular light-gathering qualities of the more expensive pair. I now blessed the extra outlay.

What had moved was a piece of wood, a long thin piece of birch, part of which was violent white in the moonlight where the bark had been peeled. It leaned up awkwardly at a shallow angle from its downstream end lying in the water, and there at the far end was my outrageous hunch. A beaver was hunched over its back legs with its tail splayed out behind and flat on the stones, and in its forefeet was a strip of peeled bark which it was eating in an attitude of apparently deep concentration. In the moonlight. Mission accomplished.

It was the clearest, the most captivating, the most memorable view of a beaver I had had, and it still is. At that point, I had been working on my beaver researches for about a year, and I was still unreconciled to the often hemmed-in, half-lit world they occupied, and more than once I had been reminded of how, at the lowest points of watching badgers, it had been forcibly impressed on me that I didn't much care for black woods at night. There had come a defining moment when I was writing *Badgers on the Highland Edge* that changed everything, for I had got lucky. And now, I had just got lucky again, and that singular beaver poised in midstream in the full glare of the all-but-full moon, peeling back a loose end of bark as if it was a banana, reached across the water and the intervening airspace and ensnared me for a willing conscript to its cause.

"I believe in God but I spell it Nature." An architect said that, the architect whose words begin this book, Frank Lloyd Wright. It has become a kind of shorthand for what passes for religion in my life. I looked up at the moon, I turned back to the beaver, I thanked that God I spell Nature. It had to be an architect, didn't it?

◎◎◎

A few days later, I looked in on the river in the early afternoon to see if anything obvious had changed in the various pockets of beaver forestry operations. In the interim, there had been more rain, more snow high up, and the river was pounding past the beaver dams. It had deepened and widened the canal, and rearranged the timber that cluttered the mouth of the pool. I had wondered if beavers could do anything at all that was remotely architectural with the river in this mood, but it had begun to look as if only a protracted freeze-up would inhibit them, and that was not about to happen on a river like this one. Its surface was particularly animated, excitedly vocal, ribbed with patterns – chevrons, stipples, arcs, swirls, white riffles, dark brown snuffles, flat patches as white as marble, others corrugated and dark as oak bark.

This winter river is a living creature, a mongrel offspring of a snake and a chameleon that swithers effortlessly from vivid blue to dark brown, gold or orange or yellow where low sun lit a last gasp of what had been a long and lingering autumn. Down by the water's edge where the beavers had been working on a group of trees since I first found this place, I could feel the redefined bank's edge tremble in the face of the river's charge. I could also see that where the beavers had felled several parallel trees into the water, the effect was to slow the pace of the river's surge along the banks, and divert ever more water further inland and further uphill in a series of new inlets.

After an hour examining the most accessible parts of Beaverland, photographing and scribbling notes and sketches, I wandered up to where a huge yew tree stands high above that long horseshoe bend in the river. The bulbous roots of the yew offer a reasonably comfortable perch with long views upstream and far over the low ground to the mountains. I could see the shingle bank, last seen in moonlight, and there, clearly visible in the binoculars, was the discarded birch branch, the exposed wood already darkening where the bark had been peeled. So the spates had not prised it free, and no beaver had returned for it, neither to peel more bark nor to deploy the timber, which would, admittedly, have involved an awkward downstream journey of a few hundred yards to the heart of Beaverland.

The image of that solitary, moonlit beaver had kept

revisiting me in the intervening days. There was something engagingly symbolic about it that demanded consideration. Sitting in the roots of the yew tree, I tried to write it down, for somehow it had crept under my skin in a way that none of my other tree-felling, log-toting, river-swimming, dam-building, sun-bathing, midstream-dawdling beaver glimpses had, not to mention the half-glimpsed blurs and tail-thrashing of beavers I had unwittingly disturbed. Yet that moonlit beaver had barely moved while I was watching it. And when it did finally move, it only walked a couple of yards to the water, stepped in and submerged. There had been no splash, no tail thrash, no lumberjack heroics, no architecture. It had left behind only an indelible memory and an abandoned piece of a partly-peeled birch branch, without which it would have been easy to conclude that I had imagined and idealised the whole thing.

And the more I thought about it, what it seemed to come down to was this: the handful of beaver landscapes I have seen are places in turmoil, in a state of fast transformation, and every beaver I have seen has been embroiled in that apparently chaotic process. It was all essentially as new to me as it was to the landscape, and the impact was all the more drastic because of its suddenness and the speed of such dramatic influence in landscapes where just a handful of years or even months before, there had been no beavers, no memory of beavers and no tradition of beavers.

But that beaver haloed in moonlight and enthroned on the shingle in the middle of the river presented an image as timeless and unlikely as El-ahrairah, the legendary and god-like super-rabbit of *Watership Down*. I surprised myself with how these few minutes had caused me to change the way

I looked at beavers. At first, I was perhaps just a little too pleased with myself for backing my own irrational hunch as an instinctive response to one set of nature's circumstances, the hunch that considered another possibility just beyond this stretch of wooded river. But every creature I have ever watched for any length of time with a view to writing it down has, sooner or later, offered up a defining moment that has justified the long hours, justified them to the writer at least. It is about being prepared to go again and again, to win a degree of familiarity with their particular landscape in the hope of becoming a part of that landscape in the mind of the creature, and watching what flows as a result.

This ten-minute audience with El-ahrairah-of-the-Beavers (I'm now stuck with my own careless image of it) revealed a perfectly painted portrait of the archetypal beaver, shorn of all tree-shadows, and aloof from the workaday turmoil of its clan. In my mind it has established itself in a kind of hierarchy of individual animals whose paths I have crossed, like the pale golden eagle I wrote about in this book's predecessor, *The Eagle's Way*, like a mute swan on a Highland loch that fell asleep in my shadow, like a Mull otter that tried to engage me in a game of hide-and-seek, like a badger boar that peed on my wellies while I was wearing them...

◎◎◎

I was still sitting in the roots of the yew tree when a light and throaty growl, a bit like a duck with laryngitis, dragged my mind back to the here and now, and before I could pinpoint its source it threw in a mildly hysterical falsetto note then the rasping creak of big wings. It was a heron. It had been hidden behind the bulk of the yew, but now appeared flying so close to me that the wingbeats were suddenly the loudest noise in my ears. It reacted to my presence with a second falsetto rasp and banked sharply away from the tree and dropped down towards the river, calling constantly to a second heron beyond the trees on the far bank. As it leaned east again directly above the river and into the trees it growled once more and then followed that with the same falsetto embellish, a sound that was followed almost imme-diately by an almighty splash. Heron knows beaver, beaver knows heron, and for the moment I'm guessing they don't get on very well.

Above: One of Scotland's European beaver *(Castor fiber)* pioneers.

Overleaf: A juvenile photographed among water lilies at Aigas.

© *Laurie Campbell*

Above: 'Beaverhandling' more food for the larder.

Opposite: This close-up reveals how the beaver can grasp the thinnest of twigs, as well as large logs for dam-building.

Overleaf: Two beavers feeding on aspen at Aigas.

© *Laurie Campbell*

Above: Stripping bark.

© *Laurie Campbell*

Chapter 7

BEAVER AND SWAN

BEAVERS DON'T WORK well in zoos. The Bute experiment demonstrated that if it demonstrated nothing else. One hundred and twenty years later, so did the restless, enterprising colonists who staged a prison break from captivity, thereby scattering to the four winds the seedbed from which the Tayside population sprung. And sometime shortly before that fateful day...

...WAAAAAHHH!!!
What the hell?!
Someone was playing the trumpet, loudly, badly. In a wildlife park.

I was standing by an enclosure in the park (a zoo in all but name), looking morosely at three beavers looking morosely at me, while someone from the park staff was telling me how much they liked apples, and did I realise I was looking at the first beaver kit to be born anywhere in Britain for 400 years. But my mind had slipped its moorings and drifted away to 1989 when I was speaking at a book event for the St Kilda Club, an event which for reasons that were never explained at the time, was held in Edinburgh Zoo, and on the way in I passed a snow leopard and knew then that I would despise all zoos for the rest of my life, so I had stopped listening to the PR speech about the beavers and the apples when...

...WAAAAAHHH!!!

"What the hell?!"

I said it out loud this time. But then I realised that the man from the wildlife park was still chatting away about beavers, and that he had either not heard the trumpeter (in which case he was stone deaf and had been conducting our interview by way of impressive lip-reading skills), or else it was a sound that was such an everyday part of his surroundings that it had long since ceased to be noteworthy. That was when I rebuked myself:

"What the hell am I doing here?"

What I was doing there was this: a friend had suggested to me that the wildlife park's shop would be a good place for me to sell a few copies of a book I had written about swans. It was called *Waters of the Wild Swan* (it still is), and after a minor flurry of book launch activity, its London publisher quickly lost interest, and to cut a long and dismal story short, rather than have several hundred copies of the first (and only) print run shredded I had bought them for about fifty pence a copy, an outlay that more or less cleaned me out. A £15 hardback reduced to fifty pence after about eighteen months – it was a harsh lesson about the meagre nature of tolerances in big London publishers. So I began looking for outlets where I might sell my stock for as much as I could get for them, and a friend had suggested the shop in the wildlife park. So that was what I was doing there. No-one was remotely interested in buying any, but as a consolation prize I was shown around by Wildlife Park Man and we had arrived at the beaver enclosure and someone or something started playing the trumpet very, very loudly and very, very, very badly. Just one cataclysmic note – WAAAAAHHH!!!

As I didn't appear to be getting through to my host by speaking in my normal and admittedly quiet voice, I grabbed his arm to get his attention and demanded loudly:

"What the hell is that?"

A slow, knowing smile crossed the face of Wildlife Park Man. He beckoned.

"This way."

We walked perhaps 100 yards, rounded a screen of some kind (a wall, a high hedge – I forget) that concealed another enclosure from the beavers' compound. At that moment, I had what remains the only proper and utterly accurate premonition of my life, one that was about to be spectacularly confirmed as we reached the end of the screen. Wildlife Park Man's lack of response to the trumpet abuse had alerted me to a new possibility, one that instantly transformed into alarming certainty. Not a trumpet player but a creature, another of the park's wretched inmates that made a noise like a trumpet... oh no, oh please God for pity's sake, NO. Oh yes, not a human trumpet player, but the brassy voice of a trumpeter swan.

No irony before or since – none since the very idea of irony was invented – was ever heavier, grimmer, more malevolent, more toxic, more calculated by all the Gods of the Wilderness the world over and all its skies and flaming underbellies, to strike coldness in my heart at the very mention of the word "zoo" forever after. Here I was, having suppressed all my troubled instincts, soothed by my friend's reassurances, blinded by the small mountain of unsold swan books that now filled every square foot of storage space in my house and other people's houses, driven by the undeniable truth that I was well on my way to going broke... here I was on a mission to sell copies of a book for which I cared passionately, and which my publisher had casually tired of in rather less time than it took me to write it. In unthinking pursuit of a few pounds I had wound up in a zoo that incarcerated trumpeter swans, those giant cruisers of the North American continent where they have symbolised fidelity and freedom and the wisdom of long-distance migration for thousands of years, those haunters of lonely nesting waters in some of the world's wildest acres. I thanked Wildlife Park Man for his time, made an excuse and left. I have rarely felt so wretched in my writing life.

Beavers in a zoo enclosure appalled me for the way the "territory" they are permitted by the zoo management is ugly and diminished and dismal, and the fragmentary illusion of a wild beaver that is on offer as it swims across its garden-centre-like pool is wrecked once you are aware of the wild animals' workaday capacity – and need – to re-landscape its surroundings, on a grand scale. But if anything, a trumpeter swan in a zoo is worse, for it can only be contained there if its wings are clipped, and its wings are the chief glory of the trumpeter swan. An eight-or-nine-feet wingspan is normal, and a mature cob can nudge ten feet. In flight they are simply the most glorious creatures in the skies of our world, and their giant stride devours epic migrations. And if you know that, then you know too that a wing-clipped trumpeter swan in a zoo, with a voice designed by nature to echo along whole mountain ranges, is a laughing stock and a freak show.

All that would have been bad enough, but an episode in my working life a few years before had ensured the boundlessness of my sense of outrage. Back in the late summer of 1998, I embarked on what remains the great wildlife adventure of my nature-writing life, three weeks in Alaska to make two radio programmes for the BBC's Natural History Unit. The opportunity came my way because of a series of well-received programmes I had made for the NHU. They were about swans. I have written about aspects of that trip several times, for it is forever drip-feeding itself into my consciousness. Until I met Alaska face to face, I had never fully understood the unfettered range of nature's daring, never fully grasped the raw and teeming limitlessness of the possibilities of landscape and wildlife when they are just allowed to be. There was so much to it that I am forever tracking back there in my mind, following the spoor of memory, as some new facet of my work along Scotland's Highland Edge or Hebridean seaboard or the eastern seaboard of childhood and youth startles me with some new recollection of moose in Denali National Park, or grizzly bears along the Kodiak Island shore, or the whale-strewn currents of Glacier Bay, or a parliament of bald eagles assembled by a carcase, or a long quiet day on a backwater of the Yukon somewhere near the Canadian border looking for trumpeter swans…

… My guides were an amiable biologist called Dave Mossop and his equally agreeable three-legged mutt, whose name escapes me now. The day was grey and damp and given to showers. The terrain was forest and bog among quiet hills surrounded by quieter mountains, and I was assured that beyond the trees we would find a small lake that reputedly held nesting trumpeter swans. I was further assured that as long as the day did not worsen, as long as the visibility did not dwindle, as long as we did not meet a bear, as long as... we would have a half-decent chance of seeing trumpeter swans. It seemed at that moment a long diversion for a half-decent chance ringed like a stockade with conditions, but a trumpeter swan expedition had been included in the itinerary at my request.

We stepped from Dave's campervan, and the vast embrace of one of the world's true wildernesses rushed down far mountainsides and near hillsides like a conspiracy of avalanches and settled tense and chaotic around us and grew still. I imagined that the only comparable circumstances on earth would be engendered by mid-ocean, mid-desert, or mid-polar ice. Into this aura we began to walk, and I remember feeling... not exactly small (which I am not) but insignificant, or perhaps "inadequate" is better. It was not the first time and nor would it be the last since I arrived in Alaska that I succumbed to the reinforced awareness that here, mine was not the species that made the rules.

The forest gathered round, Dave assured us that his dog would alert us to the proximity of bears long before we alerted ourselves while I was thinking, "Why would we not want to meet a bear?" Fear of wild animals has never been a part of my make-up (although I have met a handful of people I would run a mile from) and I admit to a degree of disappointment when I encounter it in others. The forest was unlike any other I had ever seen. I passed a tree that appeared to have been felled at knee-height, but it had been felled by no axe, nor by a passing gale. The stump ended in a rounded point like a pencil, and the "sharpened" portion was patterned all over with small indented patches. I recognised it for what it was only from a hundred different book and magazine photographs.

"Is this beaver?"

"Yep."

"Where is the trunk?"

"Broken into chunks and ferried to the lake to build lodges and dams."

"Ferried?"

"Sure. You just stepped over his canal."

"*His* canal? Whose?"

"Beaver's."

What I had taken for a patch of rainwater lodged in a rough ditch, such as you might find in any Scottish forest, was actually my first beaver-designed canal. The shore of the lake at this point was still out of sight, but now that I looked more carefully, the canal was not a "patch of rainwater" but an unbroken waterway that appeared to lead to a wide pool about fifty yards away.

"The pool?" I asked.

"Also beaver," Dave affirmed. "There's a perfectly curved dam on the far side that holds the water there, and an outlet that goes all the way down to the shore. Little fish in it too. Not that they eat fish, but plenty of things that do eat fish catch them there; thing about beavers is they're always putting food in the mouths of other creatures."

We started to encounter more and more stumps, some with felled trunks nearby or even still attached, but every trunk was stripped of branches and foliage and bark. There were also many trees that beavers had gnawed most of the way through then abandoned so that eventually they died where they stood. Dave explained that these "snags" – standing deadwood – are perfect for wood-burrowing bugs and woodpeckers; and that fallen deadwood breaks down, again over time, and increases the fertility of the forest floor, fertility that creates openings for new plants, new trees; and in the process they create new opportunities "for water to do its thing".

"They create havoc," he said, "then they leave it, and nature makes a garden of it, because nature has time."

"So the trumpeter swans have beavers for neighbours," I said.

"Yup. I guess they know each other pretty well. And sometimes, it's the other way round and the beavers have the swans for neighbours: I've heard of swans that built a nest on a beaver dam."

"Good grief! Why?"

"Dunno why. Maybe out along some of these boggy shores it's the only solid thing they can find to build on."

So that was how beavers entered my life, deep in the Yukon where Canada and Alaska seep seamlessly into each other. That was how they first commended themselves to me for their accomplishments in what is surely the ultimate form of architecture, for the architects not only design and build, but also select their own raw materials, engineer the means to move them from source to building site, and then physically move them. All that, and they are the friends of the world's largest and wildest swans.

And then there was the lake and then there were the swans. There were two adults and four dowdy, half-grown cygnets. They were a few hundred yards away and our view of them as they swam was constantly interrupted by intervening trees. Up to that point, the swans of my life had been the native mutes, Icelandic whoopers, compact little Siberian Bewicks that winter in Scotland, and occasional introduced Australian black swans. Even at distance, the trumpeters looked huge compared to all my familiar points of reference. I was wondering how close we might be able to get when Dave's dog, which had been gleefully roving ahead and around us all morning, went nuts. He yipped and growled and bounded back towards us with surprising speed for a three-legged mutt, and his short hair stood on end. Then Dave's voice:

"Holy Toledo! Bear bed."

"Bear bed?"

"Grizzly makes a day-bed where he lies up. Here."

And there on the ground was a rough circular scrape five or six feet across and a deep duvet of grasses and ferns, the most innocuous looking thing if you don't know its meaning. Once you do know its meaning, you get to thinking about who made the bed and who's been cat-napping in it. Dave knew its meaning, looked at his dog and back at the bear bed.

"I don't like this," he said.

"You think the bear is still around." It wasn't a question.

"I think the dog thinks so. Could be. Not worth taking the chance. I think a strategic withdrawal is called for."

"Bugger."

But in situations like that, you trust the expert, the local knowledge, and you defer to the survival instincts of the three-legged dog. We turned round and headed back the way we had come, back through the rain and the trees and the beaver clearings and pools, and on the way I took the trouble to notice when I stepped over the beaver canal. If it was a frustratingly unfulfilled expedition in terms of the trumpeter swans (and the bears, for that matter), it was a kind of shorthand distillation of the nature of that extraordinary country. The beavers I have been watching on a Perthshire river have kingfishers and dippers and otters and red kites for neighbours and any of these can thrill and engross me for hours at a time. The beavers on the Yukon have trumpeter swans and grizzly bears for neighbours, and unquestionably, from time to time they will be harassed by wolves.

Two days later, my producer and I flew in a small float-plane to a lake on Kodiak Island, home of reputedly the largest of all grizzly bears. As we stepped down onto the dock, a party of visitors in immaculate and brightly coloured waterproofs was boarding a second plane. They were waved off by a tall, well-built man who then turned to greet us, for we were his next clients. He was a bear guide and he was going to show us bears. He looked me up and down, taking in my spattered green jacket and richly muddied overtrousers, and he beamed.

"Ah good," he said, "dirty gear!"

I explained that we had been looking for trumpeters in the Yukon and he beamed again.

All that unfolded because I had asked my BBC producer if there would be scope on the trip to see if we could find some trumpeter swans. In the days that followed, one extraordinary wildlife spectacular piled up upon another, and we filled twenty hours of tape interviewing extraordinary people leading extraordinary lives, and it was long after I had washed the Alaskan dust from my boots that I had time to reflect on the significance of what had unfolded, and to order my thoughts into something that made sense. My life-long fascination with swans had acquired a new benchmark. I have studied a single mute swan nest in a flood-tormented reed bed on a Highland loch for more than thirty years, I have bumped into a handful of nervous little Bewick swans newly touched down from Siberia on an unexpected hill lochan folded away among quiet Scottish hills, and I have seen whooper swans nesting on a shore of black sand in the lee of an Icelandic volcano. And then there was that glimpse of a trumpeter family as far from home and as far from human habitation as I have ever been or am ever likely to be, and their utterly uncompromising wildness put a lump in my throat and a thrill in my heart, and that brief-est of glimpses of their lives had lasted no more than five minutes, and it was presaged by beavers and thwarted by a grizzly bear.

So five years later, when I arrived in a state of some trepidation at a wildlife park with a small box of swan books under my arm and met beavers and trumpeter swans fenced in and going nowhere and utterly deprived of the fulfilment of every wild instinct they possess, it is difficult for me to convey how low my spirits sank. All of which may go some way towards explaining my acute unease about the fact that one of the partners in the official Scottish Government's trial reintroduction of beavers in Argyll is an organisation that owns two zoos. Somewhere down the line, if Scotland loses its nerve about beavers and caves in to the insistent bellyaching of the usual suspects, then I worry that the temptation will always be there to listen to the siren song of the easy option. Because there *will* be teething troubles, beavers *will* build dams where farmers and landowners don't want them and beavers *will* fell trees they don't want felled. Some land managers will knee-jerk their way to the ears of Scottish Government ministers who will find themselves besieged by powerful self-interest, which has neither the time nor the inclination to await the benevolent transformation of biodiversity that will evolve as the cycle of beaver activity begins to kick in; neither the time nor the inclination to watch and wonder at the genius of nature's architect at work in our land for the first time in 400 years, because that same self-interest has shouldered nature aside and rubbished thoughtful nature conservation for more than 200 years.

Few landowners and farmers think in the long term. Fewer still are willing to acknowledge that nature may know more than they do about land management, about marshalling natural resources, about natural ways of regenerating woodland, about manipulating waterways, about using wetlands to mitigate against flood and drought, about the wisdom of biodiversity, and about the folly of monoculture. Beavers do all that. In the process, some people are bound to be inconvenienced in the short term, for their initial impact can seem drastic. Where there is goodwill, solutions can always be found, but too often in Scotland there is no goodwill when nature contrives to inconvenience land managers, and the only acceptable solutions involve shotguns, traps, poisons and dogs; witness the crude intolerance that characterises the relationship between the natural world and the wholly unnatural world of the grouse moor and the deer forest, the relationship between animals behaving naturally and people behaving unnaturally.

So in the particular case of Scottish beavers, the lingering, insinuating presence of the easy option, the zoo option, is potentially a disaster waiting to happen. Given the controversial two-pronged nature of the way in which they have come to repopulate a swathe of the country from Argyll to Angus (the right kind of beavers and the wrong kind of beavers, the Norwegian Eurasian beavers and the Bavarian Eurasian beavers, the legal and the illegal beavers, the electronically tagged beavers whose every movement is monitored and the untagged beavers whose every movement is utterly unpredictable), there may come a time when the Scottish Government is tempted to emulate the kind of Westminster Government thinking that produced the badger cull in the south-west of England in order to appease particularly strident landowning interests. A partial cull in Scotland would free the country of the illegal beavers and the legal ones could be delivered into the eager arms of zookeepers.

I don't think that is likely to happen. But until wild beavers are accommodated in the landscape and their presence enshrined in law, even the remote possibility is a worrying distraction. If you have seen beavers and trumpeter swans in Alaska whose freedom knows literally no bounds, then seen the cowed caricatures of both in a little Scottish wildlife park where they know nothing but bounds, such a fate for our briefly wild beavers is not acceptable. Surely the 21st-century Scotland with aspirations to be an enlightened and independent country in a modern Europe that has already embraced widespread beaver reintroduction, is beyond that.

Chapter 8

AIGAS:

FLOATING ISLANDS AND
WHY BEAVERS CLIMB TREES

SIR JOHN LISTER-KAYE, 8th Baronet of Grange, OBE, untypically enlightened and live-in Highland landowner, mastermind of the world-renowned Aigas Field Centre, at various times a big wheel in the RSPB and the SWT, first Highland chairman of Scottish Natural Heritage, associate of Gavin Maxwell, naturalist of distinction with a worldwide reach, nature-writer of great gifts, enthusiastic digger driver, and friend, was looking at the beaver skull in his hands with something of a wistful "Alas poor Yorick, I knew him well" affinity. It would be true, too, not that he was in the habit of naming the beavers on his loch at Aigas, but because that particular beaver would have spent its life there. He was demonstrating how far back in the skull was the hinge that released the jaw, and therefore how much power the animal could bring to bear when it went determinedly at, say, a Highland silver birch of thirty or forty summers, or for that matter any living thing with which it decided to take issue.

It is fair to say that my small circle of good friends is not over-stocked with baronets and landowners, although I did have a very good friend who was a professional dig-ger-driver. In John's case, though our paths might even-tually have crossed for the first time at some book festival or other, our friendship happened at all because of a rather remarkable letter he wrote to me. I had been dipping into a couple of his books that I have admired for years as part of the process of preparing the ground for this chapter when the letter in question fell out of the book in my hands, *Nature's Child* (Little Brown, 2004), a quite beautiful and touching book about wildlife adventures in the company of his daughter Hermione. I see that it is dated September 15th 2004. It begins:

Dear Jim, (if I may?)

NATURE WRITING

I was both horrified and delighted to read your piece in The Scots Magazine *(January 2003) on nature writing while sitting in the dentist's surgery in Inverness yesterday. I was delighted because I, too, have felt that I am an endangered species and I identified straight away with your thrust; and horrified because you thought you were on your own.* [This was a reference to the main theme of the article which bemoaned the dearth of nature writing in Scotland and suggested that when I looked in the mirror what I saw was an endangered species.]

... None of this is directly why I'm writing and sending you Nature's Child. *I am doing so to shore you up! Not only do I want you to stop thinking you are on your own in Scotland, but also to say that your own utterly enchanting little*

swan book has long been a favourite of mine – inspirationally so – and to thank you for that. But also to extend a warm welcome to you next time you are heading Beauly or Glen Affric way. There is always a bed and a bowl of gruel here for a fellow nature writer, and perhaps we should try and meet up anyway, and we can be endangered together...

I had never received such a letter before. There was such a genial big-heartedness in these few lines that I was very touched by it. I was also curiously charmed by the serendipitous nature of his encounter with my *Scots Magazine* article – in the dentist's waiting room a year-and-a-half after it was published. Our subsequent friendship grew out of that remarkable beginning, and is sustained by an absorption in nature, by nature writing (we have indeed since crossed paths – and collaborated – at book festivals), and by a shared and inspirational admiration for the writing of Gavin Maxwell. John knew him well, worked with him in the final years of his life on Eilean Bàn, and concluded the story of Maxwell's otters in his own first book, *The White Island* (Longman, 1972). My own involvement with Gavin Maxwell's writing was simply as a reader, albeit a reader who determined at the aged of around eighteen, and after reading *Ring of Bright Water* (Longman, 1960) twice in the space of a week, that that was what I wanted to do. It would take me until the age of forty to make the leap of faith that began my nature-writing life after twenty-four years as a newspaper journalist, but the seed of the whole enterprise had been the book Maxwell simply referred to as *Ring*. To this day, writing my own *Ring* remains an unfulfilled ambition, and the only target I have ever set myself. So an

Above: A beaver dam in Knapdale.

Overleaf: Trees felled by the beavers at Aigas.

© *Laurie Campbell*

Above: A beaver lodge, Aigas.

Opposite: A felled silver birch.

Overleaf: Flooding in Knapdale.

Last page: A gnawed rowan showing signs of regeneration.

invitation to partake of a bowl of Aigas gruel (it turned out to be cordon bleu gruel, courtesy of John's wife, Lucy) was not one I was about to turn down.

Back at the beaver skull in my friend's hands, people at the sharp end of beaver reintroduction work have nothing to fear other than the occasional necessity to trap and subsequently handle one. There are approved techniques, of course, but once you have seen and heard a beaver at reasonably close quarters as it systematically reconfigures a standing hardwood into so much horizontal and sectionalised timber, you can't help thinking thoughts that begin with "What if...?" You can lose a finger to an otter. I would imagine that if you were careless enough, you could lose a hand to a beaver.

John warmed to his subject with effortless enthusiasm. He cupped his hands in rough imitation of the beaver's cheek pouches, while explaining that it can suck these in so that it can chew bark and other vegetation underwater without taking in water at the same time. Beaver reintroduction into Scotland is a *cause célèbre* for him, and the loch at Aigas has been home to a pioneering demonstration project, which, he conceded, was born out of frustration. Ever since his overtures in the 1990s to Secretaries of State for Scotland Ian Lang and Michael Forsyth fell on stony ground, he embarked on a course of action with the ambition of fertilising the ground himself. To that end, he and his friend and fellow untypically enlightened and live-in landowner Paul Ramsay, travelled to various parts of Europe as far north as Norway and as far south as Spain, learning everything they could about beavers and reintroducing beavers, and how what they learned might play out in a country that had had none for centuries; learned about what constituted good beaver habitat, and about beaver reintroduction schemes that had already borne fruit. Together, he and Paul determined to establish demonstration projects, one at John's Aigas estate near Beauly in Inverness-shire and one at Paul's estate at Bamff near Alyth in east Perthshire, but not before they had scoured Scotland in search of suitable beaver terrain. There was, after all, not much point in proving that Scotland could accommodate wild beavers if there was nowhere to put them.

"So we travelled all over Scotland north of the Tay and found 111 sites of prime habitat."

I thought I had mis-heard him.

"One hundred and eleven?"

"One hundred and eleven!"

And one of them is just a few minutes' walk from the room where I was shown the beaver skull. In his book *At the Water's Edge* (Canongate, 2010), John paints its portrait and owns up to his feelings about it:

> *...I'm taking off towards a loch – my loch encircled by wild and sensuous woods. I don't mean* my *as in my car or my coat, or as decreed by a deed in some lawyer's dusty safe, but* mine *in the sense that no-one can ever steal or deny; a private, unassailable* mine, *the mine of my hopes and dreams...*
>
> *The loch is only eight acres in size. Its water is dark with peat. It is roughly heart-shaped, with one or two bays and sedgy marshes; an earth dam 60 yards long flattens the point of the heart. The burn from a smaller natural lochan flowing out across a rock sill was dammed in the late nineteenth century as a water supply, more than doubling the size of the loch.*

Tucked neatly into its own hollow, the loch is its own secret, hiding from the visible world. It nestles there on the edge of human intervention: above it the wind sings across uninhabited moorland, wild woods of downy birch, eared willow and Scots pine, rowan and aspen, goat willow and wych elm, juniper, gorse and broom crowd into its banks, and below the dam the manipulated quilting of forest and field is where people have always lived. Over the years I have created a circular trail around the loch. In places the water laps at the path's edge, which then veers off into the woods, winding over bogs and heathery knolls, only to be lured in again at a little bay or a marsh as though the walker has been drawn back, unhappy to be out of sight of the water for long. In summer water lilies burst from a surface of green plates; in winter my loch brims with pure sky.

There is a sweet ring of rightness about the fact that beavers were reintroduced into a loch, the size of which had been doubled by a human dam, a rightness of which Frank Lloyd Wright's God and mine (the one spelled Nature) would surely have approved. If they had been reintroduced instead into the original lochan, I would cheerfully have invested my life's savings (in the unlikely event that my life might ever one day know such a phenomenon) on a bet that their first act would have been to dam the outflow. That in turn set me thinking along a track I had never explored before: surely it was from the example of beavers that people first learned to build dams. And isn't it remarkable that no other creature but us has learned the beaver's skill from them, for all the countless creatures that have learned to exploit beaver architecture? We, of course, are also the creatures that took the beaver genius to its illogical conclusion, fashioning dams with such skill and daring and spectacular self-interest that (in America and China, for example) there are rivers that dam-builders have stopped in their tracks, depriving them of the birthright of all rivers, which is to reach the sea. Of all the black arts we have mastered in pursuit of the cause of perverting nature, that is surely the blackest art of them all. And when a farmer in, say, Perthshire or Angus demolishes a new beaver dam across a two-yards-wide burn he has probably already manipulated and straightened and fed with field drains, he might pause to consider the relative ambitions of beaver and man in the matter of dam-building. I don't imagine he will, but the implications are there nevertheless, in the make-up of his own species and its relationship with what we have left of the natural world.

The Aigas project began with a pair of beavers from Norway, and these were introduced into an artificial lodge John and other members of the Aigas staff had built into the edge of the loch where the path threads its way through a stand of handsome Scots pines. It's a robust structure with a removable cover in the roof. In response to the eager interest of John's two Jack Russells in whatever scent was thermaling up from somewhere beneath the cover, he removed it with great care until we could peer into its gloomy innards, but all we saw was a few pieces of wood, mostly stripped of bark, because for the last six years the beavers have used it not as a lodge but as a log store.

"They spent three days in it," said John, "before abandoning it for the opposite side of the loch where they built their own." Typical architects, I thought. And they built it of birch and willow, which both grow all around it, and neither of which grows around the artificial one. It was the first of many lessons the Aigas staff have been taught by the beavers.

As I write, the Aigas project is being run down because the beavers are beginning to eat themselves out of habitat, the inevitable consequence of containment. At this point, what John is doing by removing the beavers is replicating beaver behaviour in the wild, by which they would remove themselves to pastures new, and in the fullness of time, once the evacuated landscape has settled down to the new, beaver-adjusted reality then evolved into something else ultimately more suitable for beavers, a new generation will slip into place in the land-and-water-world of their forebears. At some point, possibly in five years, possibly longer, the loch at Aigas will host new beavers.

In the meantime, John Lister-Kaye will write a report on the past six years, which, I guarantee, will illuminate the realities and justify the rightness of a life shared with beavers. And because it will be written by one of the finest writers in the land in any genre of literature, it will be spared the turgid and mangled species of English that characterises the twilight world where dwell the bureaucrats into whose hands the fate of Scotland's beavers has been so carelessly spilled.

A long afternoon wandering round the loch with John, pausing often to inspect and discuss the visible manifestations of beaver occupancy, was a conducted tour through some of the edited highlights of what that report might contain. So we paused by the lodge, the beaver-designed one, and there was the ghost of Horace T. Martin again, wagging an index finger at me, pronouncing another warning:

The beaver lodge is generally included in the list of marvels revealed for the investigation of those who visit beaver districts, and yet no greater disappointment awaits the enquirer than the first inspection of one. Somehow the minds of all lovers of Natural History become affected by the fabulous accounts concerning this structure, and it is a shock to stand for the first time before a pile of twigs, branches and logs, heaped in disorder on a small dome of mud, and to learn that this constitutes the famous lodge. Of course the superficial glance does not convey all that can be learnt in connection with this work, but it does most completely disillusion the mind. On breaking through the upper walls, the interior... has scarcely a distinguishing characteristic...

Ah, where was I? Oh yes, so we paused by the lodge, the beaver-designed one. It was, very roughly, the size and shape of one of those three-or-four-person tents that slope smoothly down the sides and the tapering rear from a short ridge, but with a blunter appearance at the front. It was made of innumerable pieces of wood, but certainly thousands of them, and from biggish twigs to the intact trunks of long and skinny birch saplings. The "tent" was arranged so that it jutted out backwards into the water, water that was once grass-topped shoreline and shoreline-fringing trees. Nearby, a few yards from one side of the lodge, a much heftier tree had been felled so that almost its entire length lay in the water, although it was still partly attached to the shore at the root. If that is the way it was felled then it was an astonishingly precise piece of work, for its furthest end lay about a yard beyond the outermost gable end of the lodge, but I rather suspect it was manipulated in the water. However it was achieved, another slenderer tree had been felled on the far side of the lodge and arranged (I feel as sure as I can be that that is exactly what happened – the position of the trees had been "arranged") so that its further end nestled into the embrace of the bigger tree. The effect of this and sundry lesser supporting fellings was to create a lagoon of becalmed water around the three faces of the lodge that slope into the loch, and which, of course, would include the underwater entrance.

The landward area around the lodge was strewn with more or less parallel birches felled so they would lie towards the water. From a photograph I took at the time, I see that there were fifteen within about fifty yards, far too many for their parallel alignments to be a coincidence. One lay along the entire length of the visible base of the lodge, and a second had been placed on top of it and at right angles to it, so that it lay across the lodge's landward edge. A third birch trunk lay along the further side of the lodge, parallel to the first one. The idea that these formed some kind of supporting frame for the intricate architecture of the lodge was irresistible to this non-architect.

Other felled trunks lay nearby, and my guess is that some of them would eventually become lodge, for the architecture of beavers is forever a work in progress. But what you see of the lodge is rather less than what you get, for what you see is the visible eighth of the iceberg. When BBC Television's *Springwatch* visited Aigas, they installed cameras inside the lodge, and revealed an unsuspected labyrinth of passages and chambers of varying sizes. As John put it:

"It was quite a revelation to discover that the lodge is not just one big dining room where they all sit around the table eating asparagus."

The other surprise was to discover just how far the network of excavations extended inland and underground. And when you consider that in some parts of Norway and France they burrow straight into the bank with no visible lodge above ground, it rather calls into question why they need to build a lodge at all. But I have discovered that scratching my head in a state of profound bafflement is becoming an increasingly familiar behaviour trait the deeper I have immersed myself in the beaver's world. Its capacity to confound the human observer is apparently limitless. It was a subject that cropped up halfway round the loch when we were talking about that most visible aspect of beavers' presence, the gnawed tree and the peeled bark. John said:

"And they can and do climb trees."

This rather stopped me in my tracks. I was scratching my head again.

"Why would a beaver want to climb trees when it can fell them so easily?"

"That," said John, "is actually a much bigger question than you intended to ask."

"Is it?"

So he explained that so many aspects of beaver behaviour appear random, inconsistent, unpredictable. Example: when they emerge at dusk they head off in different directions from day to day. Yesterday's work is not necessarily today's work and today's not necessarily tomorrow's. Example: when they manoeuvre pieces of tree trunk or branches into the water en route to the log store or to somewhere else to eat bark or foliage, they don't take the shortest straight-line route, but often swim far out into the loch on what appear to be quite unnecessarily circuitous swimming journeys.

Example: when they go ashore to collect bracken, say, they will often go to some length to gather dry leaves, yet the first thing they do with it is take it into the lodge underwater. Example: and yes, they do apparently climb trees, just to peel bark from a piece of trunk which is out of reach when they are on the ground, and this regardless of the fact that there may be hundreds more trees to choose from.

And then, there is the matter of the how and the why and the wherefore of tree-felling technique. This has puzzled me from the outset and it still does (see the River Diaries chapters), but as with so much else, my puzzlement must surely be symptomatic of the fact that I have no experience of sustained *patterns* of beaver behaviour because those patterns are only evident over decades, and however much intimate knowledge our ancestors acquired has been obliterated from our own awareness by the sheer passage of time, by the tragic bulk of the beaverless centuries, by the disappearance under the evolutionary horizon of all those generations who could have handed down the knowledge. My own detailed observation has been on the river, as and when I can get out there. But at Aigas they have the benefit of camera traps, which can keep an eye out for what goes on at night. I was curious to know if tree-felling was always a solo effort or a group activity, or at least a paired activity. Former Aigas staff naturalist Alicia Leow-Dyke ran the day-to-day business of the beaver project at the time. She sent me the following answer for my curiosity to gnaw on:

I have only seen beavers fell trees on their own, and the way I have witnessed this is through capturing them on camera traps. I have also read that they only fell trees on their own because of the danger of becoming crushed under a felled tree. A number of deaths have been caused by a tree falling on a beaver. Luckily enough that hasn't happened at Aigas!

Although beavers fell trees on their own, it often takes them a few days/weeks/months to completely fell a tree and in that time other members of the group may have a go. Beavers often fell trees with a diameter of 10 centimetres and in theory they could fell this in 30 minutes if they kept at it, but often it will take them a few days or months, because they often move between trees.

It is tempting, of course, to go looking for a rationale that might underpin the apparently random. The harder the wood, the harder the feller must work, so move between say, a thick oak and a thin willow for a little light relief. Or: let the tougher trees weaken while they are still standing by cutting halfway through the trunk and leaving it so that it starts to destabilise (that way, time and nature assist the felling process). Or: but of course, these stabs in the dark of ignorance deny the possibility that there really is no rationale at all, because sooner or later they will fell all the trees they need to fell and quite a few that they don't need… or at least if they do need them felled, it is not obvious to you and I why they need them felled. Time, time is the secret to our understanding of nature's architect and his works. Or hers. The best way to understand a new creature in our midst is to spend time with it, to observe it on its own terms rather than ours, for ours tend not to allow for the passage of so much time before we jump to conclusions.

I also asked Alicia about the neat piles of wood chips that gather about the base of every beaver-targeted tree (unless the tree in question is so close to the water's edge that the wood chips are carried away in little flotillas by the current); do the beavers use them for anything? The answer appears to be no:

I have seen no evidence of beavers using woodchips. They have no nutritional value. I haven't seen them used in any nutritional way. They will just become another form of deadwood for the beaver habitat/ecosystem.

If that sounds undramatic, you might like to consider the following. In a good woodland, forty per cent of all wildlife – *all* wildlife – is dependent on deadwood for a living. By its very nature, deadwood impacts on the living wood slowly and diversely. A Scots pine that dies on its feet might stick around in a vertical position for a hundred years. In that time it will nurture small lives in uncountable numbers. But it only starts to feed back into the woodland floor once it has fallen over. If a living oak or beech is blown over, that process begins more or less at once, but you can imagine how long it will take such a brute mass of timber to break down naturally.

But at the other end of the scale there are beaver-generated woodchips. Mostly they are chips bitten out of the living tree, but they have been reduced to tree fragments about an inch long on average, and they do nothing with them, other than accelerate unimaginably the process of the humification phase of decomposition. The earth that succoured the seed that became the tree reclaims the tree in handy bite-sized chunks. Soil organisms get to work on what is essentially the final part of the cycle of woodland life. In the course of a single year a group of beavers on one small loch like Aigas or one stretch of river like the Earn will produce millions of woodchips, and leave them where they lie, and in the process they do the earth yet one more favour.

◉◉◉

We paused in our walk around the Aigas loch to admire a little masterpiece of beaver architecture. A tiny burn I could step over and which had fed into the loch has been diverted within around twenty yards of the shore and then dammed. But a straight dam would not have done the job because at this point the shore is more or less flat and the water would simply have spread out along the wall and flowed round the ends. So the beavers curved it. They built a horseshoe dam curving its ends back towards the hillside, and creating a wide lagoon. The dam is about twenty feet long and a metre high at the highest point. It is scrupulously built and scrupulously maintained. The whole enterprise has used nothing but burn water, wood from the surrounding trees, and endless armfuls of mud from the bottom of the loch. As a result, it looks as if it has been there forever. It is a refuge, a place to eat, and to sunbathe. In summer it dazzles with damselflies.

It reminded me a bit of a place I know well near Balquhidder in the Loch Lomond and the Trossachs National Park. Two small ponds lie on the watershed, the smaller one a few feet lower than the larger, and the two are linked by a sliver of a burn. Water lilies and reeds crowd the surface and young trees – birch, willow, rowan – stand among its thickly grassed banks. Otter tracks tunnel through the grass and in winter when the ponds freeze, you see their footprints cross the ice. I was once regaled by a keeper with the story of how he built the ponds by hand so that his boss and his boss's clients could shoot duck there. It had taken months, many many buckets and spades and barrows had been involved, and the labours of Hercules paled in comparison. If he had told me that story today instead of fifteen years ago, I would have advised him to get a pair of beavers, and maybe take a couple of weeks off while they did the job for him. Then the landowner changed, the new one had no interest in shooting, and nature has taken the ponds under its wing. One early morning I disturbed a roosting sea eagle in a tree on the bank. Beneath it were pellets the size of hand grenades.

Near the Aigas dam a small wooded platform has been built out into the loch. John leaned over the wooden railing to show me where one of the supporting posts had been gnawed clean through by a beaver. Why would it do such a thing with so many trees at its disposal? The answer was a shrug, a reference to those teeth that never stop growing, and another one to my earlier question that had prompted the response, "That's a much bigger question than you intended to ask."

John directed my attention to a tiny island of grass and other small plants just a yard or two out into the loch. I then started to notice that there were quite a few of them scattered around that part of the loch in among the leaves of water lilies. He explained that the beavers ate the water lilies, but also dived down to the bottom of the loch and disturbed the mud there to get at the water lily rhizomes, which they ate from their "hands" like a carrot. Chunks of the bed of the loch disturbed by the beavers floated to the surface, where in time, they acquired surface vegetation.

Alicia elaborated:

It took us a while to figure out what had happened when the islands had appeared, but on closer inspection we noticed that the beavers had done quite a lot of repair work to one of their lodges where they had used a lot of mud. We soon realised that the mud they had used had come from the edge and the bottom of the loch, and dislodged a fair amount of material which then floated to the surface. Over the years we have seen beavers use these islands as feeding areas and otters and mallards have also used them, and dragonflies love them. We have found evidence of otters using them as feeding areas and the mallards will sometimes roost on them overnight.

John leaned out over the railing and prodded the nearest island with his stick. It moved! The whole thing floats free, like a living raft, but as soon as he withdrew his stick it drifted back to its previous location as if it was anxious to reclaim what it had already determined was its optimum position.

There is a hide next to the artificial lodge in the pine-wood that is frequented by visitors who come specifically for a week of beaver-watching. Spring and summer are when beavers are at their most obliging for beaver-watchers, venturing boldly out into the long, light evenings. The beaver weeks are just one option from a series of themed holidays on offer, including wildlife photography, wildlife art, a variety of wildlife weeks some of which specialise in either birds or mammals, and including whisky and wildlife (thinks: must sign up for that), and botany. All of them are cushioned by the truly memorable experience of Aigas cooking and accommodation, and the company of kindred spirits.*

The point of mentioning this is the economic one, because there are those who can see no good in any new incursion into the complacent status quo that passes for the management of our countryside with its eerie echoes of the gospel according to Queen Victoria, unless that incursion can be justified by numbers. So if you must have an economic argument for beavers, here it is: 5,000 people have come to Aigas specifically to watch beavers, and in the process they have generated £3 million. And if you prefer scientific numbers instead, how about this: over the course of the Aigas beaver project, constant monitoring and testing has revealed that biodiversity within the beavers' habitat has increased by 400 per cent.

We discussed the implications of this sitting in the hide. John pointed to the interpretive material on the wall, in particular (and with the ghost of a rueful smile) to the one that says that the beavers never travel more than fifty metres from the water. At Aigas, they now know that isn't true, for their paths have since been found extending several hundred metres uphill and over rough ground. Camera traps have shown that once one beaver has made such an excursion others will follow the same path.

Sooner or later, all these things come back to the same essential truth. The only way to understand an animal is to spend time in its company. It is particularly true when the animal is a reintroduced stranger to us, like the beaver.

* www.aigas.co.uk

John Lister-Kaye's far-sighted beaver project at Aigas is profoundly persuasive. It has been something of a proving ground, and what it has proved beyond all doubt is that the company of beavers is one of nature's better ideas. That notion can be extended to encompass another one – that reintroducing it into the wild places of Britain, even after 400 years of absence, will prove to be one of the people's better ideas.

Chapter 9

THE RIVER DIARIES: III

LATE WINTER, a half-dark-at-noon kind of day that feels as if it never woke up this morning; a half-drenched-in-gloom kind of day that has the moist and torpid air of a lull between downpours. The river is more bloated than boisterous, ill-tempered with its own too-muchness. Heavy rains have wrenched heavy snow from the mountains and hurled the whole cold-tea-coloured tumult down among the beavers and their works, many of which are temporarily drowned. The river's voice is a deep and ominous throb with a frog the size of Ben Vorlich in its throat. Even in the shelter of the trees the wind still shivers with the scent of mountain snow. What could I possibly learn from such a day?

With the river in this mood, I am restricted to about 100 yards of walkable bank, walkable in the sense that for most of the way the water's tug on my wellies is neither too deep nor too threatening. At the upstream end of the beavers' canal I can feel the bank shudder beneath my feet. For the moment the canal only has one bank, the one I'm standing on. The long, thin island the beavers created by digging the canal's trench in the first place, thus severing it from the original bank, has been inundated so that its dozen-or-so standing trees wade knee-deep in the current. The pool that normally fills the bottom of a natural scoop in the riverbank is hardly recognisable as a pool now, such is the volume of water rumbling through it, this despite the fact that the pool only came into being when the canal's downstream end shovelled aside the last beaver-armfuls of mud and let in what was designed to be a quiet backwater.

At first acquaintance, the apparent chaos of felled trees here was bewildering. Bewildering because (a) I had never seen anything like it and (b) I grew up in a land with no beavers, so there were no reference points, and now suddenly... all this. The downstream extremity of the island had been under siege. Seven or eight stumps stood around, grave-stone-like memorials to their departed trunks, although few of them had departed any distance at all. Most of them fell across the mouth of the pool, helping to define and deepen it and to quieten it by deflecting away the river current, while a small, more carefully constructed dam edged out into the water from the other side of the pool, narrowing the gap, restricting the flow of water back into the river, further deepening the pool.

For the moment, only the downstream edge of the island and a few of the bigger felled trees were recognisable from that first midsummer acquaintance. Now, I looked down on it from the top of the highest bank twenty feet above the pool, sitting on one more fallen tree, staring through the half-light, doing nothing more useful than acquainting myself with the river's most doleful mood, while the barely audible calls of a flock of four different kinds of tits zinged through the trees and a pair of tree creepers explored the minute innards of the bark of an old oak that I fancied was beyond the reach and the ambition of either river or beaver.

⊚⊚⊚

What may have been an hour later, an interval during which I succumbed utterly to the mood of the day and the river, absolutely nothing had changed. I have no defence against this kind of nature-induced inertia. I don't listen to music or whistle a happy tune to fend it off. If it is your ambition to be a part of the landscape and observe the natural world from that point of view, then it follows that you should let your own mood match nature's mood. I never wrote a thing worth a damn when I was trying to fight nature's mood. Years ago now I stopped trying. *There is no substitute, even in divine imagination, for the touch of the moment, the touch of the daylight on the dream...* It is just as true when the daylight is the shade of shrouds. And suddenly there was nature's touch on the moment. And suddenly there was a head in the river.

The foreground confusion of standing and fallen timber, the swollen corrugations of the river itself, and the more or less total absence of usable light, created such a kaleidoscopic morass of shades of brown that the thing was gone almost before I knew I had seen it. But I could not have been more certain. There was a head in the river.

Then another glimpse, closer this time, a low, wide head with blunt muzzle and wide-apart eyes, a head I know very well indeed. Then there were more intervening trees. Then the head again, and then it dived. Damn.

Then it surfaced again, and in a sudden porpoising movement that lifted its tail out of the water, it identified itself very clearly. I knew then at once that it was about to come ashore and with any luck it would do so on that far edge of the pool by the small beaver dam. Instead it came ashore on that last tip of island that remained clear of the river, shook its river-slicked fur so that the whole animal stood for a moment in a bubble-shaped apparition of teeming rising and falling water droplets. When all that subsided my binoculars were full of a hefty dog otter, about twenty yards away. Now that he was out on *terra firma* (or at least as *firma* as any of the riverbank *terra* was likely to provide at that moment), he stared intently back along his spine in the direction he had come, as though something he had just seen or just encountered still held his interest. Then with tail thrust straight out behind him he deposited a spraint among the roots of one of the beaver-felled stumps. It was then that I noticed dried-out remnants of old spraints, which meant that particular patch of the end of the island was a routine stop.

The relationship between otters and beavers is as new to us as beavers are. We still know very little about it in these circumstances newly redefined by beaver reintroduction, and so for that matter do otters. Bear in mind that our beaver population stems very directly and very recently from a core of animals sourced from either Norway or Bavaria, so even their offspring bred in Britain must have at least an inherited genetic awareness of otters from those countries, quite apart from what they will pick up from day to day. Beavers would reasonably expect otters to be in landscapes like this, and know what to expect from them. But like the rest of us, our native otter population has occupied a beaverless country for centuries and may not know what to expect, any more than we do. As the beaver expands its territory, otters will be encountering it for the first time, and it is at least possible that they will have no idea what it is. I have watched many otters in Scotland in many different landscapes, and have a rough idea what to expect from them in most of the situations they might encounter in the natural scheme of things. At that moment, while I had no reason to be surprised by the appearance of that particular otter now revealed in stupendous detail in my binoculars, I had very little idea of what might unfold if it encountered a beaver, and that very little was based on secondhand accounts and scraps of reading. On the other hand, beavers have clearly been on this stretch of the Earn for quite a few years now, and in the hours of dusk, dawn and darkness, they will have had to make whatever accommodation was necessary with their otter neighbours.

⊚⊚⊚

I have not found the conventional image of a lodge on this stretch of this river, nothing like the structures at Aigas or Bamff. Instead, I have found that this large and flat-topped mound where I sit on my fallen tree, and which rises immediately on its north side from the beavers' canal and pool, has also acquired several freshly excavated holes on its west-facing slope. But these have been excavated from within (there are no spoil heaps, no tracks connecting the holes), and these suggest to me the ventilation shafts of the chambers of a large beaver lodge that has simply been burrowed into the slope, for nature's architect has not only mastered the technique of building the timber and mud fortress, but also the souterrain. Also, the base of the mound on that side is gently washed by what looks as if a dense thicket of willow and alder has invaded a moat. In the geography of this particular stretch of river, the holed slope and attendant flooded woodland is what I have come to refer to as the Centre. It was towards this curiously compelling place that the gaze of the otter on the island was magnetised, and it was from that general direction that his swimming head had first appeared.

Whatever may have just taken place, it was clear that the otter was, to say the least, unrelaxed about it. The fixed stare, the frozen posture, one forepaw raised... if I didn't think I was treading on the unstable banks of anthropomorphism, I would read into that gesture a palpable tremor of fear. After two or three unmoving minutes, the otter stepped off into the quieter water on the canal side of the island, dived, and reappeared within seconds in the middle of the pool. There it made a few leisurely circuits before diving again then resurfacing within seconds, and this time clutching a smallish brown trout by the head. It returned with the fish to the island spit, devoured it, slipped into the water and began to swim slowly up the canal, only to perform a swift U-turn in its own length on the surface, and return to the pool where it performed more circuits, none of them leisurely.

I turned the glasses away for a few seconds to look up the length of the canal, and just where a ragged edge of river water spilled over a low and very old stone wall there was a second swimming head, the head of a beaver. This, I told myself, is getting ridiculous. It is (a quick check) one o'clock in the afternoon, and there is a beaver apparently patrolling the entrance into the Centre and the end of the canal, shuttling backwards and forwards like a sentry. I wondered if the unusually niggardly level of daylight accounted for this, not that the beaver was deluded into thinking it was evening, but rather that it judged such low light constituted a threshold of safety. Or perhaps it was just that the otter had been nosing around in the floodwaters below the burrowed-out hill, and that was justification enough to make a pointed declaration of occupancy.

I turned back to look for the otter in time to see it emerge from pool to river, and disappear downstream in some haste. Within minutes the beaver was gone too and following that very unsatisfying conclusion I was left to contemplate the deepening gloom of the afternoon and in a state of equally deep mystification. It turns out I am not alone.

◎◎◎

The word is "commensalism" – an association of two species whereby one species obtains benefits from the association without the other species being harmed. That at least is the dictionary definition. And although the word has been applied to beavers and otters, initial scrutiny of the scant evidence out there suggests there is one word missing from the end of the definition – "much". In the 2004 bulletin of the International Union for the Conservation of Nature's Otter Specialists Group (OSG), Bob Arnbreck wrote, *A commensal relationship between otters and beavers has been documented, and use of beaver ponds for foraging and beaver lodges for denning is regarded as characteristic behaviour of otters…*

So far, so commensal, but then, … *Reports of beaver hair in otter scats suggests possible otter predation on beavers.* Uh-huh. The report was based on a study of Canadian beavers in an area with five beaver colonies in up-state New York. It noted otters foraging in beaver ponds, making latrines, scent mounds, rolling areas and dens around ponds and on beaver lodges. There were observations of beavers swimming round otters while they fished. But *generally the animals did not interact.* In a six-year study, otters were seen on a beaver lodge forty-six times, but only six encounters between beaver and otter were observed.

But in the OSG's 2008 bulletin, Daniel Gallant and Andrew L. Sheldon observed, … *the relationship may not be the commensal one suggested by some researchers.* This followed *a rare observation of reciprocal aggression between two river otters and a beaver* in which bites were exchanged. The paper urged that *future research should aim to determine the frequency of these agonistic* (another word I have filed away for its crossword puzzle potential) *events and their overall cost to beaver survival and reproductive success, to determine if this relationship is really a commensal one or some integration of strong positive and negative effects.*

The Canadian beaver and the North American river otter are admittedly different species from ours, and you could be forgiven for suggesting that there is so much more in Canada and the northern United States that is more truly wild than there is here, but nothing in what I have heard and read suggests there is much difference in the behaviour of the species on either side of the Atlantic. But I am amazed that the relationship doesn't come up more often in the scientific and popular literature that has considered beaver life and lore. The beaver's natural predators are generally listed as wolverines and bears and occasionally wolves (not to mention the unnatural one, which is ourselves and our associated consequences – our dogs, our vehicles, our agricultural chemicals...). But otters as predators somehow seem to have managed to stay under the radar most of the time in most beaver landscapes in Europe and North America, and this despite the fact that in addition to being a constant neighbour of many beavers, the otter is also more than capable of getting into a beaver lodge by the underwater door.

And then John Lister-Kaye told me on our wander round his beaver loch at Aigas that "they don't seem to get on very well with otters", and described several stand-offs, and one occasion when "three otters charged a beaver". The beaver eluded the charge and disappeared underwater. He also told me that otters visit the lodge, and wondered if they perhaps shared it, or were sizing up the possibility of killing the kits. A contribution to the Aigas blog in May 2011 went into more detail:

Two otters were patrolling the water's edge around the new beaver lodge in which we believe this year's kits are already born. At first we thought they were fishing for small brown trout, but it quickly became apparent that they had another purpose. We did not see them catch fish at all, although later, when they left the area of the lodge they quickly caught fish and went ashore to eat them.

Their mode of patrol was constantly diving in the area of the underwater lodge entrance, spending a little time on the surface, but when they did surface they were almost always a few feet from the lodge. Underwater they were only travelling up to 20 yards before surfacing again, turning round and quartering the same area again.

The beavers were late in emerging and not until half an hour after the otters had left the loch. By the time the light failed they had only seen one sub-adult beaver and *it looked more cautious than normal.*

John's analysis of this behaviour was that the otters had probably heard young kits yelping inside the lodge and saw an opportunity for predation. Their behaviour, he suggested, was *more than just patrolling, but perhaps lacked the courage to actually attempt an entry into the lodge.* It was the first time such behaviour had been recorded at Aigas.

In past years, he wrote, *we have seen the very young kits emerge in early June on their own, bobbing along at the edge of the lodge. If one had been so foolish as to do this with the otters there, I am sure they would have killed it.*

Two days later, Alicia Leow-Dyke noted that during a guided visit to the beaver hide, guests had been watching two beavers swimming in the loch, when an adult otter appeared on the bank, entered the loch, and began swimming beside them, ... *but this was much to the displeasure of the beavers.* The otter seemed unperturbed and eventually returned to the bank and disappeared.

☉☉☉

I left the river that grim late winter day colder and wetter than any of my previous beaver explorations had managed to achieve, but with my mind as deluged with the

endeavour's possibilities as its riverbanks were with melt-
water. Somehow the inconclusive and intoxicating otter
close-up and the brief, gloomy, ghostly blur of the distant
beaver patrolling its stronghold, had intensified the nature
of my quest in Beaverland. That quest was – and still is –
simply to try and get under the skin of a well-established
Scottish beaver landscape (and it is just possible that this is
the oldest established of all, considering that the Earn was
the river where the accidental beaver adventure began), to
try and persuade it to give up some of its secrets, to try and
understand it a little, to try and write it down. Slowly –
everything about the enterprise was achieved slowly – I was
becoming more convinced that I had stumbled by accident
into what was surely a particularly significant terrain in the
short history of beaver reintroduction. I knew that it must
have begun here or hereabouts. I also knew that there were
more beavers four or five miles downstream, I knew by
now that they had also travelled upstream all the way into
Loch Earn, I knew that they had at least explored into the
wider Trossachs area's eastern ramparts – the River Balvaig,
Loch Lubnaig (and its significant satellite in my awakening
to the beaver's cause, Lochan Buidhe), the River Teith,
and at least one of its tributaries near Callander, and I sur-
mised that it was more than likely that particular exploration had originated from here, and had voyaged by way of
Loch Earn and south via Glen Ample into Loch Lubnaig,
a voyage that crucially transported the Tayside beaver population into the catchment area of the Forth, and a whole
new realm of possibilities. They were now, for example,
only one watershed removed from the newly established
Great Trossachs Forest, and what could be more relevant

to that enterprise than the presence of the beaver, the great forest manipulator? Meanwhile, east and downstream along the Earn was how the beaver had originally found the Tay, and extended its reach north into Angus and east Perthshire, and thence inexorably deeper into heartland Highlands and nudging towards the west. This very stretch of the Earn, which I had only chosen to explore on the basis of a chance conversation, could well be the pivotal hub of all that. And now the enigmatic presence of otters had just swum into that heady mix.

I have purposely remained aloof from the official beaver trial in Argyll, because it is too contrived and too bureau-cratically overloaded for my taste and my nature writer's instincts. But given its geographical location so close to the West Highland seaboard, I thought it must have acquired some insight into the relationship between beavers and otters in its five years, in which case its official report to the Scottish Government should have something useful to say on the matter.

I grimaced at the tone of the Scottish Beaver Trial's website*, with its promises of family fun and the chance for children to play at being a "beaver detective"... Yes, and if you see a beaver kit with its head accidentally wedged inside the jaws of a playful otter, put your hand over your child's eyes so he/she doesn't see what happens next...

The final report is written in a species of English in which I am not fluent (and have no desire to be) but the section on beaver behaviour does indeed contain references to pre-dation on kits which I have judiciously edited in your best

* www.scottishbeavers.org.uk

interests, removing references to cadavers, for example: ... *Two kits were predated: one died as a result of traumatic injuries to the head, possibly caused by a large predator. SBT staff thought it might be a domestic dog, a diagnosis strengthened by comparison with a kit killed by a fox.* It later transpires that the fox predation is also an assumption. The report also notes that of four or five kits born in 2012–13, only two were observed the following spring... *It is most likely they are deceased (though the likely cause remains unknown).*

And that's it. Otters are never mentioned. It certainly cannot be the case that there are no otters in the area. I have seen them several times over the years on the River Add just to the north of the trial area, on Loch Sween just to the south, and on the coast at Kilberry. It is simply unthinkable that they do not frequent the waters of the trial area. And it would seem, from experience elsewhere, that they are attracted by the presence of beavers, to the extent that they turn up with some regularity *on* beaver lodges and fishing in beaver pools. It is possible, of course, that interaction between the two has simply not been observed, but it strikes me as curious that yet again, an awareness of otters as a predator of beavers has slipped under the radar. And this, remember, is the official report of a five-year trial, the findings of which would have been the basis the Scottish Government used to reach its decision.

I am wondering if there is a wider problem. Otters were transformed from vermin into wildlife superstars by a single book, Gavin Maxwell's *Ring of Bright Water*, published in 1960, never out of print, and now rightly revered as both a classic of nature writing and a classic of literature. Among many people whose lives were acutely affected by that book,

I have already indicated that I can number two in particular who are relevant to the matter in hand – John Lister-Kaye and myself. It would be futile to deny that either of us is immune to the allure of otters.

People love otters. They are high on the must-see list of many visitors to Scotland, and increasingly to those otter outposts in England and Wales. If we see them eating anything at all, it tends to be fish and crabs, and that doesn't trouble us. Otters are heroes of the natural world, and perhaps there is a widespread subconscious reluctance to publicise the fact they can also be villains... *reports of beaver hair in otter scats suggests possible otter predation on beavers.*

There are good reasons why otters would be attracted to the vicinity of beaver lodges, and it is the same reason that the *British Columbia Magazine* included in that 2013 report on the benefits of beavers: *Fishery managers used to destroy beaver dams because they thought they interfered with salmon migration. Now they're learning that beavers actually improve life for salmon and other fish. Beaver ponds provide excellent rearing habitat for juvenile salmon...* And you can be sure that if fishery managers have begun to tumble to the wisdom of encouraging beavers on their waterways, then otters tumbled to it, oh, quite a few thousand years ago. And if that has always been the primary lure, the secondary lure is surely the availability of young beavers as food, and very young beavers in particular. I am not at all sure that the simple presence of beaver hair in otter scats is proof of anything; it could just mean that during a confrontation between adults of both species armed to the teeth, you might say, an otter took a lump out of a beaver. Did anyone check the beaver scats for otter hair? The record is silent.

Otters actually entering the lodge underwater is a tricky proposition in which there is no guarantee the otter would come off best, but otters encountering kits on the surface in June in circumstances described by John Lister-Kaye would amount to an easy meal. And it would not necessarily be a particularly endearing spectacle for the growing legion of beaver aficionados to witness. And the legions *are* growing, witness the number of people who book a week at Aigas specifically to see beavers; witness the enthusiastic endorsement of hotels and guesthouses in Argyll and Tayside. Beavers are proving to be good for business as well as good for biodiversity. Might there be, then, a reluctance to tar the otter with the same brush as, say, the domestic dog or the fox? Perhaps one of the tests of these next few years will be our willingness to tell it like it is.

Chapter 10

BAMFF:

FOOTPRINTS IN THE SNOW

We are very concerned that beavers are spreading widely and in a completely uncontrolled and illegal way.

– Drew McFarlane-Slack, Scottish Land and Estates, responding to a YouTube clip of a beaver swimming in the River Leny, near Callander, Perthshire, 2012.

Sightings like this show us how beavers would naturally move and disperse around the Scottish landscape, seeking out the very best habitats first before filling the gaps in between. It is very exciting to come across a wild beaver in a completely new area when they have been extinct for 400 years. For those lucky enough to find one, you will remember it for the rest of your life.

– Louise Ramsay, Scottish Wild Beaver Group, responding to the same clip.

IF THE WILD beavers of Scotland (also known as the Tayside beavers, the illegal beavers and the wrong kind of beavers) should ever stage a coup to seek retribution against what they might call (with arguably a degree of justification) the wrong kind of landowners, they would first establish their safe haven, capital city and strategic headquarters on the Bamff estate just north of the east Perthshire town of Alyth, just where that fair county begins to nudge steadily uphill and northwards out of the Lowlands and towards the indisputably Highland heights of Glen Isla and Glen Shee. In fact, and without staging a coup as yet (beavers tend more to a live-and-let-live, turn-the-other-cheek-pouch disposition), they are already in the throes of the safe haven/capital city project, not least because from a beaver point of view Bamff happens to be in the hands of the right kind of landowners.

Paul and Louise Ramsay acquired their first beavers in 2002, sometime after Paul and John Lister-Kaye had scoured the land and much of western Europe in search of habitat and knowledge and returned with a very good grasp of both. Not long afterwards, I had visited Bamff to talk beavers as part of a radio programme I was making. At the time, my sum total of practical experience of beavers was that Alaskan woodland adventure, en route to find trumpeter swans, and that old conversation with Don MacCaskill, so less than exhaustive. Furthermore, my reading about beavers had been strictly transatlantic. But that day when I met the Ramsays rekindled something that had lain dormant since that other day five or six years before deep in the Yukon when beavers had first commended themselves to me for their accomplishments as architects and for establishing themselves in my mind as the friends of the world's largest and wildest swans. Bamff fanned that particular spark again, and something new began.

The details are hazy now, but I remember Paul showing me a pond or a lochan where the beavers had been installed, and where they had almost immediately dammed the outflow, which had the effect of doubling the size of the watersheet. This was my first Scottish beaver dam. Memory suggests it was only a couple of yards across. It didn't look much, but what did impress me was the sheer bulk of some of the pieces of wood the beavers had incorporated, and I suspect that I used the word "manhandled" at the time. Now, I won't give the word houseroom, and I won't let anyone else use it in my presence when the subject is beaver architecture, because what they do is so much more accomplished than the m-word implies. Remember Ernest Thompson Seton's praiseworthy assessment: *The Beaver is the animal that most manifests intelligence by its works, forestalls man in much of his best construction, and amazes us by the well-considered labour of its hands.* So the appropriate word is beaverhandle. It does not exist but it should (and perhaps as the beaver finally spreads far and wide across the face of the land once more, it will), because it is the only verb that can ever do justice to the unique nature of their architecture.

So in a sense, the slow-burning adventure that finally led to this book began that day in the Yukon, but was rekindled that other day a dozen years ago at Bamff by the Ramsays' patient, hospitable, indomitable, single-minded zeal. Without them, and the complementary endeavour of the Lister-Kayes at Aigas, the future of beavers in Scotland would have been far less certain than it is, far fewer people would know anything about them, and their cause would be bereft of outgoing, far-reaching, forceful, persuasive champions. And it is no coincidence that when the wild Tayside beaver population began to consolidate its presence from those unplanned beginnings (the first confirmed sighting was on the River Earn in May 2001, seen by eight people in two canoes), began to explore the vast and beaverless potential of the Tay catchment, they should turn up at Bamff. The original Bamff beavers were contained within their territory by an electric fence, but they did not break out into the wide world beyond, the beavers from the wide world beyond broke into Bamff, and for very good reasons. The best of these is that nature knows when it is on to a good thing, and never declines new opportunities to prosper.

Beavers are in the process of relandscaping Bamff, and relandscaping is another word we may have to get used to, along with beaverhandle. When I returned in the winter of 2014–15, I found the place transformed, or at least transforming. The best way I can think of to describe a walk through the beaver landscape of Bamff, as it begins after thirteen years to hint at the nature of things to come, is as an exclamation of joy. Because this is not a theoretical exercise arrived at by professional career conservationists using computer modelling and giving due consideration to the consultation process with all the relevant stakeholders (to deploy – just this once – the wretched buzzword of choice), reconciling differences into insipid compromise. This is flat-out, full-on, uncompromising nature at work. This is nature gesticulating meaningfully with outspread arms, inviting us to watch and learn, exhorting us to think big and think wildly, teaching us by example that nature is better at this than we are. It is not "tidy". It is not "pretty". But nor is it "wanton destruction" (and all these are the basis of observations I have had thrown in my direction as if the responsibility for the nature of beaver architecture was mine). It is wild. It is a means to an end, and that end is that it will only ever get wilder.

The only single innovation that could possibly offer a greater service to nature is to reintroduce wolves, and given that Scotland is, as yet and alas, psychologically unprepared for such a leap of faith, there is nothing at all that can compare with the reintroduction of beavers. And surely a successful beaver reintroduction will pave the way for a successful wolf reintroduction.

On a spectacular day of uninterrupted sunlight on frozen snow and frozen beaver-distributed waterways, and in the company of Paul and Louise and Polly Pullar (our mutual friend and nature photographer and writer of distinction), we walked in a landscape such as Scotland has not witnessed for many centuries. The popularly accepted extinction date of 400 years ago won't do here, because extinction is inevitably preceded by a last protracted and desperate struggle to survive among a threadbare population scattered into isolated vestiges, and in such circumstances, anything like natural tribal behaviour ceases to exist. But 21st-century Bamff demonstrates the *beginnings* of beaver-led transformation, the like of which can only have happened before on very few occasions since the last Ice Age. Perhaps climatic upheavals or disease may have caused temporary lapses across the last 10,000 years, and the beaver has briefly disappeared, but those beaver footprints frozen into the snow a few hundred yards from the Ramsays' front door are nothing less than the spoor of history, the imprint of a timewarp, symbolic of the march of the beavers of all time across the land. I was rather taken aback by the fact that the sight of them put a lump in my throat, and not for the first significant time in my life, the words of Norman MacCaig pushed themselves to the front of my mind. They belong properly to a sequence of poems he wrote on the death of his great friend in Assynt, Angus MacLeod, but so powerful is their imagery that I have summoned them again and again when the occasion seemed to demand it, and here was one such:

> He is gone, but you can see
> his tracks still, in the snow of the world.

Those snow-and-ice beaver tracks intrigued us all. We gathered round them like archaeologists at Brodgar, reading ancient runes whose meaning was vividly and lucidly rendered. But when we stepped carefully round them and walked on and the conversation wandered away from them too, I had trouble getting them out of my mind. I knew of course that they could only have been there a couple of days at the most, and might easily have been the work of the night before we had found their trail, yet the memory it engenders is of an encounter with permanence, symbolic of the beavers of all our eras, and waiting in patient incarceration to be discovered, so that we might shed daylight on their living history. My mind drifted, a trifle unnaturally perhaps, to the discovery of Ötzi, the 5,000-year-old Ice Man of the Italian-Austrian border country in the South Tyrol, and his still functional yew bow and arrows; this was not that, but if they'd found his footprints instead of his deep-frozen body...

I suppose the point is that these new generations of beavers at Bamff are tampering directly with our idea of the history of the land, the *natural* history of the land, and these footprints are their signature. They are now numerous enough here to constitute something like a naturally established community in that their numbers are not monitored or controlled or the subject of scientific evaluation lest they exceed some bureaucrat's idea of what is acceptable. And they have been here longer than almost anywhere else in Scotland, long enough for their new landscape to begin to look old because they are working it with the same age-old technology of every generation of their forebears.

And, of course, mine is an illusory mindset in this regard, because only the beginning of the process is evident. Perhaps our grandchildren or their children will be the first generation to witness at first hand the first completion of that cycle, the first to acquire the awareness of its teeming benefits, the first to recognise the true significance of what was set in motion in the early years of the century. But the first symptoms of opportunity and recovery for wildness are already becoming apparent.

Paul had mentioned the arrival of a small flock of snipe that had settled a few days ago in the wet flushes that had blossomed near a beaver dam, and the wader of a different hue that was keeping them company. Suddenly a tight wedge of seven birds raced away from us, low across the land, then rose into a high arc that doubled back behind us and vanished, while an eighth bird, solitary and low-flying and flashing a vivid white rear at us, headed off alone in the opposite direction – a greenshank.

Beaver activity became more and more evident the

further we walked, until we reached what looked like a kind of heartland where a dam more than a hundred yards long had made a miniature Flow Country of an area fed only by springs, so that standing water – or rather standing ice – a few feet or a few inches deep, from puddles and ponds to lochans and canals, had me reaching back to Alaska in my mind. And as prevalent as the glimmers of sunlit sheen on slate-grey ice were the prostrate forms of felled trees; the dying-on-their-feet half-felled trees that still awaited the final, fateful bite that would tip them beyond the point of no return; the standing deadwood, some of which pre-dates the beavers and some where beavers had stripped the bark low down but never bothered to fell the tree; trees too mighty to be felled that have nevertheless been assailed by the rest-less attentions of would-be-architects who bit off a bit more than they could chew in the form of an ankle-height ban-dana of stripped bark; the sculpture park of roundly finished knee-high stumps, some pointed straight at the sky, others doubly rounded like breasts (the comparison is as obvious as it is inevitable) where trees that had double trunks had been felled one trunk at a time, and stumps still locked in a smoothly-turned twist with the fallen trunk from which it has declined to part company so that the trunk lies at an acute angle and the ultimate reach of its height is now sub-merged in ice and lower than the root that spawned it and to which it is still miraculously attached. All that amounts to the crazy paving of nature's architect at work and in search of his unending and insatiable quest for the raw materials he needs to fulfil his art and fill his belly.

In the midst of so much rearrangement of the woodland furniture, there was one tree that stood in such a state of spectacular and incomprehensible dishevelment that it had already acquired a monumental aspect – a monolithic tribute to the heroic nature of the beavers' trailblazing endeavours here, but a tribute that had gone so horribly wrong in the execution that it might also stand for the eccentric, improvisational nature (it's like jazz, you know) of the beaver's art. And immediately, I recognise that the phrase "one tree that stood" does not do justice to whatever unfolded, because most of the tree no longer stands, including the part that is now ten feet off the ground.

And even now, studying Polly's photograph of it showing the rest of us gathered around the obelisk of its sundering, it is difficult to reconstruct the events that could have resulted in its final configuration. Before the beavers had turned their attention to the tree, it was an apparently healthy birch about 35–40 feet tall, and with a trunk about fifteen inches in diameter at the base. Crucially, the trunk forked about six feet off the ground, a consideration that inevitably contributed to what happened. The beavers had eaten about halfway through on one side and had obviously made some inroads from the opposite side when something extraordinary happened, and how I would love to know whether there was a beaver at work at the time, and how it had responded when the unthinkable happened. I fervently hope it lived to tell the tale.

The trunk suddenly split right up the middle from the beaver-gnawed portion to the fork and several feet up each limb above the fork until gravity kicked in and the two elements of the redefined tree parted company. That meant that all that remained standing was a plinth about two feet high below where the beavers had been at work and a forked length of half-a-trunk about ten feet high, and looking weirdly like a giant catapult. That left the rest of the tree – the split half of the trunk, and pretty well everything above the fork, so another thirty feet of tree and the entire spread of branches – and its inevitable course, which was to crash to the ground, but it was not so simple as that. The forked part impaled itself as it fell on one of the spikes of the standing catapult, which in turn had the effect of hoisting the severed end of the trunk ten feet in the air, brandishing its beaver pencil-point credentials at the sky like a rocket ready for lift-off. At that moment, what had been the top of the tree now came to rest in a tumult of snapping branches and cracking, creaking limbs, and there the whole improbable improvised sculpture acquired its new form, and the beavers of Bamff had their monument.

☉☉☉

On such a day, snow and ice are vivid storytellers. The snow said the beavers went this way, a pine-marten was here (so they must know each other, and what secret knowledge might pass between them on a starry night… or do they go out of their way to avoid each other, their mutual suspicions unconfirmed?), a heron walked here – but gingerly – to the edge of the ice and then took off, and mallards harbouring no such doubts about the roadworthiness of the ice just kept on walking past the heron's last footprints and stepped out onto the ice and left a drunken trail of snow-rimmed web prints behind them for a few giddy, chaotic yards before they negotiated a felled tree trunk by squirming beneath it, and stepped from the ice onto the next square yard of *terra firma*.

Then the ice took up the story. The ice said that the beavers had not been intimidated or subdued by the ceiling of ice that had domed and closed in their underwater world, and that they had found ways through it into the overworld. And how it told that particular story was as follows:

Here is a big pond where the water has been purposefully backed up and widened behind a beaver dam on a hill burn. And here in the middle of the pool are four thicknesses of ice, four shades of grey. There is firstly the pervasive thickest ice that cloaks almost the whole pool from shore to shore, and that is a pale, silvery shade of grey. It is also patched with old frozen snow, and it is streaked with the going-somewhere-important short-cut diagonals of a fox. Or foxes. But this topcoat of ice accommodates a pool-within-a-pond, a long, narrow, and slightly rounded area of thinner ice. Its rim shows as a hard blue-black edge, above which rises the two-inch-high cliff of extra thickness of pond ice, a silky-smooth wall. Open water lingered longer here, sustained in the first days of the cold weather by the beavers until the constant ice-breaking became too arduous as the freeze deepened, and all that pool ice is a darker, steelier shade of grey.

Within the pool-within-a-pond are two other shades. One is almost white, a patch no more than two feet by three feet, and that signifies where beavers kept open a surfacing hole, a bit like ringed seals in the Arctic ice, albeit without the attendant problem of polar bears. But the quickening freeze had sealed that too. Finally there is an oval-shaped opening of no ice at all, and that is the pool-within-a-pool-within-a-pond, and its shade echoes the thick pond ice, except that because it is open water it catches something of the whiteness of low afternoon sunlight and that softens the shade of grey to something a little more homely.

This oval is the most intriguing part of the ice's story. It could accommodate (my hasty guess) two swimming beavers, perhaps three at a push. I was intrigued to see that the surface was half full of brown leaves, and then noticed a further little spillage of the same leaves just where the frozen pool met the thick ice of the pond. A trail led across the ice directly to that little heap of leaves, and from there it turned sharp left across a yard of the thin pool ice, before disappearing at the edge of the open water. How recently, I wondered, did a beaver cart a spray of winter-dried leaves from its food store across the ice and into the open water, and from there into the underwater entrance to its dining quarters?

So snow and ice are vivid storytellers on such a day. Sometimes, though, they leave it to your best guesses and your imagination to supply the ending.

Paul Ramsay had pointed out a detail in the dam at the tail of the pool, the first of several dams like canal locks (surely another civil engineering concept we borrowed from beavers) where the burn drops down a shallow valley. It had been breached when a spate practically burst it in two. It had been a straight dam like the others, but now it isn't. Instead of rebuilding the dam where it was breached, they rebuilt it behind the breach and in the shape of a vee that reaches two yards back into the pond, so that the next spate will divide and the water shed away from the vee to either side. So nature's architect knows about buttresses, and I can hear in my head Frank Gehry's quiet transatlantic voice (half Canada, half California) articulating his approval:

"Liquid architecture. It's like jazz – you improvise, you work together, you play off each other, you make something."

⊙⊙⊙

I learned a new word as I explored the possibilities of Bamff's spectacular harvest of deadwood: saproxylic. It means dependent on deadwood, and it applies to uncountable numbers of invertebrates, and I suppose, indirectly, to the species that prey on them – saproxylic by proxy. That conversation with Alicia at Aigas about the piles of woodchips that gather about the base of a tree during the beavers' felling operations had mobilised a train of thought that delved into the unglamorous terrain that deadwood occupies in any anthology of wilderness. It would take an exceptional speaker to rivet an audience with tales of nutrient cycling, or of the slower-than-sloth-slow but steady-as-a-rock release of nitrogen, slow and steady being the necessary characteristics of naturally occurring nitrogen production to maintain the wellbeing of the soil. The deterioration of deadwood is what fires up the process. It is also hard to overstate the importance of its role in storing carbon, which is on page one of every primary school textbook on greenhouse gases, or at least it would be if anyone was ever brave enough to write one. I'm thinking *The Green Gruffalo* might do the trick.

The sheer quantities of fallen timber in a beaver landscape may look like an elephant's graveyard or the aftermath of a chainsaw massacre, but what you are really seeing is the first unpromising-looking moments of a phenomenon that given time will evolve into a process of constant woodland regeneration that revitalises pretty well everything, a thing of natural grace and beauty. This is the caterpillar phase that begets the chrysalis that begets the butterfly, the ugly duckling that begets the swan. All we have to do is to learn to wait and resist the temptation to tidy it up.

Beaver-felled trees, the stumps of beaver-felled trees, the standing but halfway-gnawed-through trees, standing but dead trees where root systems have been inundated by permanent beaver-induced ponds (not alders though – they love the water treatment; it's why they thrive along riverbanks and lochsides, and why they were the timber of choice for the piles that supported crannogs 3,000 years ago), and trees that are neither standing nor felled but recline at an angle to the vertical and are still tenuously attached to their roots because their progress towards the horizontal was halted by the snagging branches of its neighbours... all these collectively and individually signify the means by which the sustained presence of beavers launches a tsunami of perpetual regeneration across the face of the land. That's why biologists and silviculturists who have studied woodlands with beavers as a fundamental driver of the ecosystem can say with confidence that forty per cent of woodland wildlife is sustained by deadwood, including a third of our woodland birds, two thirds of our bat species, and that eye-wateringly long list of invertebrates. I was pondering a form of words to articulate their place in nature's scheme of things when I discovered that John Lister-Kaye had just addressed it and contrived a beautiful and succinct solution in his book *Gods of the Morning* (Canongate, 2015), so I quote it gratefully:

It is the invertebrates that trouble me. They are the uncountable legions from which whole food chains build, eking energy directly or indirectly from the great universal gift of carbohydrate that surrounds us, and passing it on to a myriad higher organisms. Without the bugs the swallow can't skim the summer skies; the brown trout, the otter and the osprey would fade away; the rooks would fail and the goshawk would vanish forever into the dark woods.

Walking through the woods at Bamff is to become aware of taking a step back and a step forward at the same time. The return of the beaver is inevitably attended by the sense of historical time when the land without the beaver was unthinkable. But there are few more forward-thinking gestures that our species can make in the reality of the 21st century than to give back to nature something that belongs to it, something which our ancestors stole from it. One of many consequences is that the gesture offers a moment of reappraisal for the very land itself and everything that uses it. The land has very little experience of having lost opportunities handed back to it, and the opportunities implicit in beaver reintroduction are more or less limitless. Listening to Paul and Louise as we walked, and later over a leisurely lunch, it suddenly struck me how they did not come across as land *owners* but rather as people who lived within the community of the land. As soon as the notion lodged in my mind, a second one nudged it aside: where had I heard this before? The community of the land was a phrase that fell too naturally into place... then I remembered. Aldo Leopold.

Leopold's book, *A Sand County Almanac* (OUP, New York, 1949), which has never been out of print, is the undisputed worldwide nature writing masterpiece of masterpieces. I have not read all the world's books of nature writing, but among the many I have read, nothing else has ever got close. Its truths just go on getting truer, and it is astonishingly well written. And in its foreword, there is this:

Conservation is getting nowhere because it is incompatible with our Abrahamic concept of land. We abuse land because we regard it as a commodity belonging to us. When we see land as a community to which we belong, we may begin to use it with love and respect. There is no other way for land to survive the impact of mechanised man, nor for us to reap from it the aesthetic harvest it is capable, under science, of contributing to culture.

That land is a community is the basic concept of ecology, but that land is to be loved and respected is an extension of ethics. That land yields a cultural harvest is a fact long known, but latterly often forgotten.

And much later in the book he returns to the theme, and in the process he provides the answer to all those critics of Scottish beaver reintroduction who demand, "What's the point?" Conservation, he wrote, is a state of harmony between men and land, and I'm sure that if I had uttered that line out loud in the Bamff kitchen, Louise and Polly would have added a pointed "and women" after the word "men". Or even before it. Leopold continued:

By land is meant all of the things on, over, or in the earth. Harmony with the land is like harmony with a friend; you cannot cherish his right hand and chop off his left... The land is one organism. Its parts, like our own parts, compete with each other and co-operate with each other. The competitions are as much a part of the inner workings as the co-operations. You can regulate them – cautiously – but not abolish them... The last word in ignorance is the man who says of an animal or plant: "What good is it?" If the land mechanism as a whole is good, then every part is good, whether we understand it or not...

And really, it is not at all hard to understand why the beaver is good. Something utterly remarkable is at work at Bamff. The beaver is back from the dead, and the community of the land has a spring in its step.

Chapter 11

NEW HAMPSHIRE:
THE SWAMPWALKER

Other than humankind, no animal plays a greater role in determining the fate of wetlands than the beaver.

– David M. Carroll, *Swampwalker's Journal – A Wetland Year* (Houghton Mifflin, New York, 1999)

At first glance, it's easy to see why people might think a beaver's busy work is not so beneficial. And yet in places... where climate change is causing snow in the mountains to melt earlier in the year and droughts to last longer, beavers and their dams are the equivalent of a finger in the dyke. They can't prevent climate change, obviously, but they can take the edge off some of its effects.

– *Defenders of Wildlife* magazine (Washington D.C., 2010)

THE CRUCIAL THING that we lack in our efforts to create a climate of understanding about the place of beavers in Britain's ecological make-up is cultural memory. It was rendered extinct centuries ago, every bit as extinct as the beaver itself. Unlike the beaver itself, cultural memory cannot just be reintroduced and then go to work immediately. Nor can it be restored or reclaimed, we have to build a new one, and that can only be achieved by living through the passage of time and absorbing what the beaver has to teach us.

Beavers embody a kind of timelessness based on the cyclical nature of their behaviour that deals in multiples of decades, centuries even. But in 21st-century Britain – and alas for beavers – we do not. Ours is an impatient society that puts a price and a limited time frame on almost everything. If our government announces a plan to reintroduce a previously exterminated creature like, say, the sea eagle, the lynx, the brown bear, the wolf or the beaver, our society wants to know three things:

How much will it cost?

How long will it take?

And what the hell's the point?

No matter that increasing numbers of people are willing to pay good money and travel long distances for the chance of seeing beavers in the wild in Britain, they represent a tiny proportion of society as a whole, and society as a whole already believes it knows the answers to its own questions:

Too much. A waste of money.

Too long. A waste of time.

There is no point.

Adding to the sheer tonnage of widespread disinterest, there is the stuck-in-the-mud mindset of much of landowning society that is instinctively hostile to the slightest ripple on the surface of those Victorian backwaters where so much of it still wallows, still pays unquestioning, slavish obeisance to the two sacred cows that cloud its judgment and its pretence at a modern vision – the grouse moor and the deer forest. And at the first ripple on the surface of the first beaver-induced puddle in the first field-edge, much of farming society starts singing from the same hymn sheet as landowner society, the same weary dirges aimed at stifling nature conservation initiatives at birth.

And all this is at least partly a consequence of the fact that we lack a cultural memory of living with beavers. We lack a living resource that can eloquently articulate the self-evident truth about beavers, which is that, *given time*, they are remarkably accomplished ambassadors for the natural order of things. If only we could unearth a human witness who was born into that cultural memory, who adopted it for his own inheritance, who became a lifelong student of wetlands himself, who dedicated his working life (and a fair bit of his childhood) through his writing and his painting to spreading the gospel of wetlands in the face of innumerable threats to their very existence, and whose work is so inspiring that it has been recognised with a remarkable national honour – say, for example, the MacArthur Foundation's American Genius Award.

As it happens, I know where to find such a person – Warner, New Hampshire, so far north up the eastern seaboard of the United States that it borders the Canadian province of Quebec. His name is David M. Carroll, of whom Annie Dillard once remarked, "David Carroll is a genius, a madman, a national treasure", which is an eighteen-carat recommendation in my book. And as it also happens, our paths have crossed, albeit at a distance determined by the east and west shores of the Atlantic Ocean. And we have a mutual friend, Sherry Palmer, a New Hampshire landscape painter of genius herself. She introduced David and me by post, and we remain loosely in touch by post and email. Sherry still lives in New Hampshire, but she and I share an addiction for the Isle of Skye, and it was there that we met at a 70th birthday party for our mutual friend, Andrew Currie. Andy is no longer with us. He died aged eighty-four while I was writing *The Eagle's Way* and that book is dedicated to him. He was a brilliant naturalist and he and David Carroll are, I suspect, birds of a feather. Sherry still travels addictively to Skye to paint every year, and I still travel addictively to Skye to... well, just to be on Skye. By such happy accidents kindred spirits find ways to diminish distance and where our paths cross lies life-affirming common ground.

When I began writing this book and first pondered the dilemma of the absence of a cultural memory, and wondered what I could do about it, how I could overcome the stumbling block it seemed to present, I suddenly thought of David Carroll. There is no-one like him in Britain. In fact, his like is not possible in Britain because of the huge beaverless gulf that long ago consumed any shred of our awareness of the beaver in our historic landscape. But if it were possible to time-travel and glimpse the cultural memory our distant ancestors used to carry in their heads, it would look like the one to which David Carroll can still lay claim. So I re-read the three of David's books I already owned, and I found what I was looking for, the solution to the dilemma. And when I contacted him for permission to quote from them, not only was it generously given, but he also insisted on sending me the two books of his that I didn't have, and so a package of my own recent work passed it somewhere in mid-Atlantic going the other way, and our established connection was strengthened.

What I found in David's work was not just the evidence of a cultural memory from the perspective of one who has both inherited it and is constantly adding to it so that it will live on after him, brightened, invigorated, and enriched. It was also that he has written it down with the vividness of a writer who is also a painter. His books are illustrated by his own meticulous drawings and extraordinary jewel-bright paintings, but the writing too is visually rich. Here he is setting out his stall in his introduction to *Swampwalker's Journal*:

It is my delight and good fortune to have spent a large measure of my life in wetlands. For close to five decades now, from an intuitive boyhood bonding to a more scientific perspective in later years, I have moved among vernal pools, marshes, floodplains, and peatlands. Later science has done nothing to diminish earlier poetry: answers only unlock questions, and specific knowledge only deepens the mystery of the earth's landscape and life. As a boy, I knew ponds and swamps and streams; if the word "wetland" had been coined, it was of rare or specialized usage, and I never heard it. Today one can hardly glance at a newspaper or a television program without encountering the term. Magnificent even in their present broken and besieged state, wetlands have become arenas of intense human debate.

... Though all wetlands are tragically diminished and under incredible pressure in the human-serving modern landscape, what they hold can still be found, often surprisingly close at hand. Moments outside the human world in the shallows of a

marsh... will bring intimations of the spirit that moves with the water, the light, and the life of the marsh.

David's particular passion within the teeming tribes of wet-land is turtles, a passion he conveys with an elegance that is utterly infectious, so that I found myself longing for a Scotland awash with turtles. But in his book, *The Year of the Turtle* (St Martin's Press, New York, 1996) he acknowl-edges the turtle's debt to nature's architect:

I approach the back end of the marsh and make my way to the beaver dam, a remarkable construction that has altered the land-scape and made possible much of this world of turtles. Built across a drainage channel originally cut by an outlet brook, it turned a small, glacial pond of several acres into a wetland covering scores of acres. The strategic placement of this thirty-yard embankment of interwoven branches and mud replaced a forest with a marsh and provided a habitat for an entirely different spectrum of plants and animals. At this time of year, the water level of the marsh is even with the top of the three-foot-high dam – small spillways run over it to fall into the outlet brook below. When I descend into this shallow brook, I can crouch down and look out across the surface of the water being held back, with a turtle's-eye view of its plane. I wade through the icy runoff, which cuts a clear, narrow stream over white sand, bordered by deep-green banks of moss and arbors of overhanging alders. Ferns are just beginning to thrust out of the thawing earth.

Re-reading David Carroll from the new perspective of a beaver novice (and no matter how deeply I may immerse myself in its world for the purposes of writing this book,

it is undeniable that a beaver novice is what I am), I recognised at once the quality that is missing from anything
at all that has been written in my own country – intimacy;
and in particular, intimacy born of a lifetime's immersion in
the beaver's wetland world, and of a lifetime that dovetails
seamlessly with generations of past lifetimes of swampwalk-
ers and swamp creatures. There is continuity from the past
to inform the present, and that enriches knowledge in ways
we can only admire from a distance, and I pray that with
the passage of enough time we and our heirs will acquire a
knowledge as rich and articulate it with a comparable grace
and generosity. As *Swampwalker's Journal* advances through
its wetland year the spoor of that particular strain of intimacy becomes more and more visible. Here we are in May,
and we are introduced at once to a new phenomenon that
simply does not exist here yet – "long abandoned beaver
ponds". Read, watch, listen and learn:

*I follow a stream course to a series of long-abandoned beaver
ponds. The dams that created these impoundments have been
eroded over time by water and ice and by what Robert Frost
termed "the slow, smokeless burning of decay". Deep-muck
sediments collect in the lingering ruins of the beavers' work,
and here wetland gardens grow. The afternoon air is scented
by the leaves of wild mint I crush in traversing the bottom pil-
ings of a dam. Spring-blue and brightly golden-eyed, uncoiling
wands of forget-me-not, another plant that finds a niche on
woodland beaver dams, edge the green mint clusters. Beaver
dams that have fallen into desuetude become profuse botanical
gardens of wetland plants not adapted to flooded ponds or the
saturated muck of drained ones.*

In the first beaver pond the water is shallow, a little less than two feet at the deepest, and clear. It escapes in murmuring rushes and whispering trickles over and through the ancient dam. If the beaver were still active here, the sound of this much water rushing through their dam would quickly bring them to the site with branches, root clumps, mud plaster, stones, and other implements of repair. With legendary engineering, beavers maintain a precise and constant water level, usually two to three feet deep, gauging it to lie even with the floorings of their lodges at all times, in all seasons. In doing so, they stabilize the water level over extensive wetland areas...

... the natural cycle of beaver dams is being played out: after the supply of preferred food plants – aspen, willow, birch, and maple – ran out, the beavers had to move on. Duckweed is beginning the division upon division that will cover much of this pond, and the floating leaves of yellow and white water lilies expand on the surface to begin their claim on the sun-flooded water. Every shift of my wading stick sends green frog and bullfrog tadpoles darting in all directions. The muck is so soft that they sink into it a bit when they settle on the bottom. It is too shallow here for beavers, but painted turtles, finders and keepers of ponds in almost any state imaginable, stay on.

There is little standing water in the next pond upstream, which was abandoned earlier. Downed trees lie in a tall wet meadow of bluejoint reedgrass, joe-pye weed, boneset, spotted jewelweed, wool-grass, fringed sedge, and soft rush. A number of persistent trunks stand here and there, bereft of the spreading crowns of dead branches that once held great blue heron nests, but persevering as bleached pillars riddled with entrance openings of tree swallows, woodpeckers, and other cavity nesters. The dam is almost entirely breached, and the stream run *that had become lost in the broadening depths of the pond is once more cutting its way through the sediments of this gradually draining beaver basin. After the beavers abandoned this dead-tree swamp, the dam eroded in stages, and its impoundment became first deep marsh, then shallower marsh and shrub swamp. The extensive backwaters converted to wet meadow as the level of standing water steadily dropped. Over time, as the final impediments to stream flow give way, drained wet meadow will convert to upland old-field, forest will replace old-field, and a redefined woodland-bordered brook will run where acres of ponded water once stood. And then one day a two-year-old beaver, who voluntarily left or was forcefully driven from his parents' lodge and pond, will journey up the brook. His bright eyes will judge the re-established food supplies and the topographical setting to be right, and he will begin cutting saplings and dragging them to the streaming water, setting them in place and packing them with mud. The cycle will begin again. Impounded trees will drown, and great blue herons will come back to establish a rookery. A fringing of shoreline marsh will arise anew; water lilies, pondweeds, and duckweeds will spread great beds of floating leaves over the surface of the reclaimed pond as aquatic plants seed in or sprout from long-dormant seeds. Ducks will swim and turtles will bask where white-pine forest had stood for a time…*

Cultural memory, you see. That's what's missing. And – in time, in time – that is what ours will look like.

I allowed myself a smile of recognition at one in particular of David Carroll's observations in *Swampwalker's Journal*. It was this:

With their judicious upland riparian cuttings and their far more
extensive workings in riverine and wetland riparian habitats,
beavers can be thought of as consummate wildlife managers.
The term "wildlife management", often used in the environ-
mental polemics of the day in reference to human manipula-
tions, is an oxymoron. We should have learned long ago to
simply leave the proper natural space, to respectfully withdraw
and let wildlife manage wildlife.

Let wildlife manage wildlife. It is a recurrent theme of my
own writing ever since *A High and Lonely Place* (Jonathan
Cape, 1990). Twenty-five years later, while some of my
ideas have changed and others evolved, it remains the
unshakeable foundation stone on which the architecture of
the layers of my core philosophy has been constructed. In
Scotland more than most places, we have demonstrated that
in the matter of wildlife management the track record of our
species is one of far-reaching and spectacular incompetence.
The Victorians ushered in new perversions and depravities
and achieved the nadir, and evidence of the chill hand they
brought to bear on nature still pervades the air in the 21st
century, still poisons the land with its prejudices, and still
calls it wildlife management.

One prevalent symptom of such a negative inheritance
is that even conservation thinking is often hamstrung by
caution and timidity. Our attempts at reintroduction are
controlled with computer-age technologies that fit aerials
and transmitters and collars and wing tags to anything and
everything that we cannot bear to let out of our sight for
fear that it causes distress to another member of our own
species or bites the hand of a tourist, or worse, eats a grouse

chick. The grim irony at work here is that the outrages and oblivions visited by our forebears on those same species over centuries play no part in shaping what we are pleased to call conservation policies.

All of this is nowhere more glaringly visible than in this twin-track beaver reintroduction, the official trial with its vast budget and equally vast bureaucracy and its overarching need to control, and the unofficial one, which has proceeded at nature's pace and with nature's freedom. Farmers and estate owners in some parts of Angus and Perthshire have responded to such rashness on nature's part by demolishing beaver dams. At least one has felled trees on the bank of a burn along the edge of a field so that the beavers could not, which must rank as one of the most wrong-headed innovations in the history of Scottish land management. Some farmers have protested at the burrowing of beavers into flood defence banks built by people to drain land and channel water in a certain direction to achieve a particular effect. Which creature in nature does that most remind you of? And again, any sense of irony about how beaver-like that sounds is quite absent.

David Carroll has a passage in *Swampwalker's Journal* that may provide some pause for thought on the subject. If you have doubts about the rightness of "let wildlife manage wildlife" as a first commandment, this may go some way to assuaging them:

Along the populated sections of this river, where beaver dams are repeatedly destroyed and the animals themselves shot or trapped, they may take to living in the riverbanks in response to human measures to eliminate them. A species of beaver native to Europe that had been trapped to near-extinction abandoned dam-building completely and instead dug lodges in riverbanks. When, after many years, legislation was enacted to protect these animals, they resumed their historic way of life almost at once, building dams and lodges on their impoundments. It was as though the beavers had representatives in the halls of legislatures, who reported the good news to them. I can report no such good news to the beavers on this river. There are no assurances for any of the animals and plants that live along it. Among other things, a golf course has been proposed that would take a mile of the river bank in a town downstream, cutting it clear to the very edge of the water. With ever-increasing economic pressures and demands for places for humans to live, work, and play in the region, critical habitat and riparian buffer zones could vanish in less than a decade along the whole length of this river, as it has along so many brooks, streams, and rivers throughout the country.

But here, for now, habitat abounds. I could pass within a couple of yards of a black bear or moose and never know it. Beavers appear to be entrenched and in charge, something I take as one of the best signs in a wetland ecosystem.

Cultural memory is a source of light, an illumination cast by the blazed trail of ancestry, but its light only reaches us if the trail is unbroken. In this particular endeavour, our trail that dovetailed with beavers was obliterated beyond any possibility of repair. So this new relationship with beavers, such as it is, has been quite untouched by anything remotely approaching cultural memory. It is still much too soon, the relationship is still in its infancy, and as such is generating more heat than light. A handful of people, notably pioneers like John Lister-Kaye and Paul Ramsay, have begun to get a feel for what life with beavers could be like, but in terms of cultural memory, we are all in the same boat. We can study the experiences of relationships in other countries and learn from that, but our own unique environment, our own unique history of land use and wildlife intolerance, and that centuries-long beaverless gulf stand like a dam that holds back the necessary understanding that is a prerequisite for thoughtful conservation.

So we have to start from scratch with beavers. We will have to do it again with wolves when the time comes (and it *will* come). So in many ways, cramming the trial into five years then thrusting a decision on Scottish Government ministers based on that leaden-footed report is a hopelessly flawed process. The time to consider the evidence should have been not at the end of the first five years, but rather the first fifty. By then, we will know what a dam looks like when it has succumbed to "the slow, smokeless burning of decay", and what happens when it dematerialises, by then we and the beavers will have lived through an entire cycle of beaver colonisation. By then we will have witnessed every stage of the beaver cycle in the landscape, and by then we will be in a position to assess the true nature of the beaver's long-term impact on bio-diversity, on salmon and trout rivers, on wetland, on woodland, and on the limitless opportunities it creates for other life forms. And by then we will have a sense of what our own cultural memory will look like, and we will have something to pass on down the line to our grandchildren and theirs. By then, we – and the beavers – will have learned to adapt and accommodate each other in our very different ways of working with the land and the waterways. And by then, we may finally have evolved a new relationship with the land ourselves that is more generous towards nature, that has abolished Victorian landscape practice and prejudice, that is willing to let wild-life manage wildlife.

Chapter 12

THE RIVER DIARIES: IV

There are no empty hours in these wild places, no unit of time in which nothing happens. There are durations in which it might appear that nothing has changed. But something is always taking place. For how long now have I observed no more than the shadow of the pine in its incremental shifting as constant, if not as continually observable, as the glimmering water drifting by? There is the invisible passage of time, revealed by the sundial of this white pine. I am so aware of this place, this crossroads of life and the seasons, as a theatre of time. There is as much time coming as passing... it flows over me as the nearby water flows over a fallen alder stem or as the pine tree's shadow moves over the earth. Do I dream the day or experience it? Watch it go by or go with it?

– David M. Carroll, *Following the Water – A Hydromancer's Notebook* (Houghton Mifflin Harcourt, Boston, New York, 2009)

THERE ARE TIMES when I return to the river after an absence of days, occasionally weeks, look around at the now familiar alignment of felled trees and partly felled trees, of the dams on the far bank, of the canal and the pool and its half-finished (as I see it) architecture and think that nothing has changed. Then I start to get my eye in and new things fall into place.

One of the problems of mostly watching during the daylight hours a landscape that is mostly worked at night, is that you see not the beavers' labours themselves but the results of the labours. A loch would be easier than this tree-shrouded stretch of river (which is *very* dark at night), more open, more watchable in moonlight, free from the obliteration of spates, more rewarding for the evening shifts. But these are the nearest I have to "local" beavers, and, it is unarguably a place that is undergoing spectacular transformation.

It began with this little group of trees. The first tree I saw was more substantial than I had expected to be a legitimate beaver target, a birch about three feet in circumference where the gnawing had begun. I thought they would never get through it. Nine months later, they never have. They got to the halfway point in no time at all, and made a token dent on the opposite side. Then they left it. In my personal geography of the site, I still refer to it as the First Tree. In the next few months they felled five adjacent trees, all of them slenderer, including one that had two slender trunks from a much wider base. This they tackled in a completely different way, beginning almost at the base of the tree, cutting a ring all the way round and cutting a tapering profile, again all the way round, until they reached the point at which the trunk diverged into two. They then began to work on both trunks. The result of their work at this point was a sculpture that looked remarkably like bared buttocks, so I began to refer in my notebooks to "the Buttock Tree". Both trunks fell within a few days of each other. One was gnawed cleanly and must have fallen with the last beaver bite and a hasty retreat to safety. The other was left standing but held in place only by a section of un-gnawed trunk about an inch wide. Within two days the trunk had given up the unequal struggle and simply split. When it fell, a vivid white and jagged pinnacle of broken tree remained attached to the bare buttocks. But the tree had fallen into the water and perfectly parallel to the others. I now recognise this as one of a variety of felling techniques. The trunk is cut until what is left is so narrow that eventually the sheer weight of the tree will effectively fell itself.

Between these two – the First Tree and the Buttock Tree – another double-trunked tree was felled, again with a jagged spike for the sculptor's last flourish. The second trunk to fall lies with its landward end held a yard up in the air, because in falling it lodged in the vee of the rump still embedded in the bank. But the other trunk from the same tree is missing. My guess is that it is over there, sixty yards away on the far bank and slightly downstream and wedged into the base of a dam. It has been moved intact, and it is about thirty feet long. Did one beaver do that?

◉◉◉

I have a noticed a strange thing. The shallow water immediately below this group of trees is no longer shallow. The beavers appear to have been digging mud. Presumably they are swimming armfuls of it across to the dams on the far bank (there are none on this side of the river, only uncompleted gestures on the canal apparently long abandoned, and that inconclusive outer edge of the pool which is strewn with fallen trees, but nothing that could be construed as a dam). The thought occurs: are they building a second canal, less than 100 yards downstream from the first?

◎◎◎

The river beaver's life must be very different from beavers on a loch, or in a situation like Bamff. A ferocious spell of prolonged rains, big winds, and heavy snow on the mountains made a monster of the river. It drowned the dams on the far bank and swirled around the roots of the trees in the beavers' principal area of forestry operations. The canal had become part of the mainstream, and the "island" the beavers had created when they built the canal was simply invisible. A few marooned trees marked the spot. What do they do in the face of this kind of onslaught? I made three visits at the height of the deluge. Each time I thought the river could not possibly get higher. Each time, it did. Farting against thunder is a wise undertaking with practical possibilities compared to repairing a dam when what is left of the entire structure is several feet underwater, water moving at a ferocious rate of knots and pounding the banks like the sea at Cape Wrath. But I am more convinced than ever now that the main lodge is in the bank with the ventilation shafts. Nothing else makes sense, or rather nothing else makes sense to me.

◎◎◎

About the possibility of a second canal: at the height of these storms, the water beside the Buttock Tree shore was noticeably more tranquil than the mainstream because it is downstream and in the lee of the flat-topped mound that juts out into the river and accommodates the lodge and the pool and provides one bank of the first canal. So by deepening the water in this quiet shore, are they preparing the ground for a new purpose-built lodge? And am I beginning to think a little bit like a beaver, or is that fanciful nonsense reflecting the fact that I am completely out of my depth?

◉◉◉

The rains stopped in mid-March. I gave it another two weeks to settle down. When I went back, it was to find the river as docile as a slug. It was also lower than at any time since the previous summer, and the changed nature of the banks suggested that an entire regiment of beavers had been deployed to rectify the storm damage. The first thing I noticed, however, was that the First Tree still stood, and appeared to be untouched in the interim. Next to it the four parallel felled trunks still lie, but whereas their further ends used to lie in the water, they now lie on a long, narrow, shallow-domed island of mud that I had simply never seen before. At once I remembered the islands in the loch at Aigas that suddenly appeared, and which were subsequently discovered to be excavated from the bed of the loch by the beavers and had floated up to the surface. But the island I was looking at now was about three times the size of anything I saw at Aigas. I am as sure as I can be (a frail qualification for the degree of certainty at work here) that there are rather more beavers here than at Aigas, where the numbers were controlled by farming out the juveniles each year to other beaver projects. Yet I have only had one glimpse of two animals together and otherwise never seen more than one at a time, but I suspect there are two or perhaps three groups, and I would guess there could be around a dozen animals altogether. With the long, light evenings of spring ahead of me, my ambition is to convert some of that guess-work into actuality.

Theory: the beavers have scooped tons of mud from the riverbed here, and carted it to the far bank in armfuls, for dam-building, of which more in a moment. In the process they have created the possibility of the second canal, but a canal is only a canal if it has two banks. The sudden appearance of the long, low island parallel to the bank effectively creates the canal by presenting itself as another bank, at which point a new thought surfaces (in the manner of a new floating island, I thought brightly). Is this lagoon-like area of tranquil water in the lee of the big mound actually a beaver-created lagoon, and is that out there – that "new" island – actually the line of the original bank? Has the combination of the falling river level and the underwater engineering of the mud excavators raised the height of what was already there (like a reef of mud still anchored to the riverbed by the root systems of old trees and other vegetation) and have the beavers laboured to consolidate it as a new canal bank? In which case, is there a plan for a new lodge on this shore?

About the dam-building on the far shore: at the high water mark of the storms, the three dams were drowned. Whether anything had survived of them or not I have no way of knowing. But now, it is very clear that all three have been rebuilt. It is also clear that they look more substantial than before, although that could be a consequence of the lower water level. Two of the dams are parallel to the shore and buttressed against the press of the river by fallen trees. The third one is at right angles to the bank, and therefore much more vulnerable. In its new post-storm configuration, two large sections of bark-stripped trunks have been deployed at the base. None of these dams is very high, about three feet at the most. Each of them looks as if it is designed simply to create an area of still water next to the bank. So are they sheltering entrances to food stores, or perhaps another bank lodge? Two of them are very like that first dam I saw on the river near Callander, but more substantial.

Question: so if the construction work is on the far bank, why the particularly laborious task of carting the mud from this bank? Two answers spring to mind. One is that they don't want to deepen the water on the far bank, and they do want to deepen it over here because they are canal building. The other is that the work is easier here in the quiet water. A third consideration hovers around in my mind, namely that the idea of a particularly laborious task does not enter into the equation. A beaver's life is, by definition, one long particularly laborious task, and besides, it is a life apparently characterised by a complete absence of the phenomenon that you and I would recognise as logic.

And speaking of logic, the obvious conclusion to draw from what I have just witnessed of the rise and fall of the river is that bank lodges make much more sense here than built lodges of sticks and mud and stones.

◉◉◉

The beavers have sculpted a bald eagle. It stands erect on the riverbank near the upstream entrance to the original canal, in the midst of another area of major timber operations. In a chaotic, tight little group of trees, most of which have been felled over quite a long time (the different shades of the bared wood suggest months have passed between the first felling and the most recent), one stump is particularly conspicuous. It is first of all two-tone, so the upper part was worked on later than the lower part. The top of the stump is more rounded than the normal pencil-point stub, and as a further embellishment it looks as if the sculptor abandoned the project in mid-bite, because one bite-sized and bite-shaped chunk is only partly severed from the stump, leaving a shape that contributes a beak and an eye to the bald eagle. The two-tone nature of the sculpture, pale above and dark below, completes the graven image. Two other stumps, one on either side, lend a slightly less convincing approximation of two half-raised wings, so that it looks as if the thing is about to take off.

This sparked off another train of thought. Is it possible that there is an aesthetic element to at least some of the beaver's tree-felling activity? Every now and again, I have come across a situation in which the felled wood – either the trunk or the stump – has been more elaborately worked than was strictly necessary to get the job done. Here, for example, is a grassy stretch of bank, a rare thing along the beaver's domain. It has had three trees, all of which have been felled, all three perfectly symmetrically pointed. Not far away there are two slim trees that have leaned out over the water. Both have been felled in an identical pattern, bitten to the point where only a thin sliver of trunk connects the upper part of the tree to the stump. Both trees have then split and fallen towards the water, leaving matching pointed splinters. To all intents and purposes they are identical twin stumps. Accident or design?

In the same part of the riverbank, a felled trunk with a perfectly rounded pencil-point end has been further embellished since it was felled. It is completely stripped of bark and a yard or so from the end, it has been eaten into so that it is narrower for several inches, a perfectly rounded section a few inches long that reduces the trunk to about one third of its width. It is frankly beautifully done, but what on earth is the point? Aesthetics, or just one more example of the beaver's chaotic and inconsistent approach to logic?

☉☉☉

The pool, the canal, and the island have been most transformed by the abrupt change in the river level, not least because all three are restored to the beavers' landscape after inundation by the river, and the beavers have lost no time going to work to renovate their various projects. It looks to me as if they have been most active in the pool. The damming of the pool is much more advanced. There have been substantial additions of felled timber across the mouth of the pool from the island, and more, smaller-scale work from the far side, where the previous beginnings of a dam had been swept away in the storms. None of this looks like the more formal dams on the far bank of the river, but rather as if it might be a loose framework of obstacles designed merely to slow the passage of water out of the pool and back into the river. If the beavers were to dam the whole thing, it would be a much longer dam than anything else on this part of the river, and it may be that they have already achieved what they wanted here, the flow of quiet water through the canal into the pool and then back into the mainstream, now that their required depth of the pool has been established. It remains to be seen if the construction of canal and pool is a prelude to a new lodge. If they have a built lodge in mind rather than another bank lodge, there is now no quieter part of the riverbank than this, and it is quiet because they have made it that way.

◎◎◎

The low level of water at the entrance to the canal has allowed me to get over onto the island for the first time. It is wider than it looks from the riverbank, and the whole island is given over to beaver operations. Their tracks are everywhere, and on the outer edge in particular, many more trees have been felled than is apparent from the riverbank. Given that most of these would have fallen directly into fast-flowing water, they have either been floated across the river to the dams or round the end of the island into the mouth of the pool, or they have escaped completely and are now somewhere near Perth.

The middle of the island is chaotic with broken trees, felled trees that rot where they lie, brambles, and other progress-fankling bushes. There is a constant stream of tits, finches and treecreepers through all this, a testament to the natural food supply, for which the beavers must take much of the credit.

You get a much better appreciation of the canal from the island. It looks as if it has been there forever, an absent-minded little backwater of the river. But it wasn't here twenty years ago and the island was riverbank, and the sheer slog of beaver labour has re-landscaped everything. Trees have been felled here and there across the canal, and what I think of as half-hearted dams have been built, and these seem designed only to slow and deepen the flow of water through the canal. But the resultant shaded pools have many visitors, and nothing appreciates the flow of small fish through the canal quite like the kingfishers. There is a thin, high-pitched "chee-ky" cry, a vivid, zipping blur low over the water and an abrupt hand-brake turn, and there perched on one of those half-dams is the bird. Somehow, the confines of the canal intensify the impact of a creature already endowed with tropical splendour. The stance is erect, the colour the brightest thing in all Perthshire at that moment, the jewelled jauntiness offset (and slightly diminished) by a bill like a *sgian-dubh*, and the thing is still for all of ten seconds before it plunges, blade first, into a pool that might have been designed specifically with kingfishers in mind. It is back where it was in ten more seconds and begins to beat the brains out of a small fish on a piece of beaver dam which has been beaver-cut at both ends, and beaver-stripped of bark along its entire length. It was a happy day for the local kingfishers when beavers moved in.

The darkest part of the canal is just before it emerges into the pool, and is quite out of sight from the riverbank because of the steepness of the high, flat-topped mound at that point. From the island, though, its secret is uncovered. Several small trees have been felled into the canal and

these too have their grateful visitors. I was edging pains-
takingly through the middle of the island, making slow
and, I suppose, noisier progress than I would have liked,
when the depths of that dark and furtive water resounded
with a splash, a cry like one of Tam o' Shanter's skirling
hags, a second splash, and the rasp of huge grey wings. The
voice and the wings were easy enough to identify: a heron a
dozen yards away creating enough turbulence to rise almost
vertically from a standing start and negotiating a morass of
fallen and standing trees is hardly likely to be mistaken for
anything else. But what about those splashes?

I have heard quite a few beaver alarms now, and seen
some of them in execution, the tail brought into ferocious
contact with the surface of the water, but this didn't sound
quite so alarmist as that. My first thought was small beavers. I moved as quickly as I could to improve my view
of the pool and the river immediately beyond it, but no
beaver stirred, and as it was still mid-afternoon, I had hardly
expected it. But the source of the splashes suddenly revealed
itself – or rather themselves – out in midstream. Four goo-
sanders – two ducks and two drakes – had obviously been
working the depths of the canal, and two of them had been
perched on the fallen trees. The hospitable generosity of
beavers just goes on giving and knows no bounds.

Chapter 13

RETURN TO LOCHAN BUIDHE

THE LOCHAN LIES in its flattish, grassy, reedy basin, a jewel in the open palm of the land, or an ever-wakeful eye watching the sky for swans, a beckoning glimmer of light and moonlight. You see it best from the forestry track high above its western shore. Between track and shoreline there is a steep slope of hoary old lichen-draped oakwood with furtive, secreted clusters of juniper bushes in its midst. Two months from now, you can walk here knee-deep in acres of wild hyacinths and drunk on the scent after rain. At the foot of the slope, where the Callander-to-Oban railway used to run (closed in 1965, thanks again, Dr Beeching), there is a long, skinny, dead-straight birch wood. Nature's preferred option for a disused Highland railway line is almost always a birch wood. Beyond the birches, the land is two-thirds of the way to wetland, into which willows and especially alders have waded. Beavers, when they finally take it into their heads to colonise here and realise Don MacCaskill's old dream, will supply the missing third.

Then there is Lochan Buidhe, the Yellow Lochan. If that Gaelic word *buidhe* troubles you, it sounds a little like "boohee", as long as you don't overdo the "h" in the middle, but rather just hint at its presence, as it is in almost every Gaelic vowel. It is possible that whoever named it had a sense of humour or was afflicted by colour-blindness. On a day like this one, right on the cusp of winter-into-spring with April only a week away, it is never yellow whatever else it may be. It is dark silver in sunlight, it is dark pewter when the sleet showers tumble round the hillside from Balquhidder's mountainous glen to the north, and when the showers have clattered on down to Loch Lubnaig just to the south and the northern sky is blue again, the lochan is a saturated shade of the deepest royal blue. Of course it doesn't last, not on a day like this one, because over the space of four hours the pattern of shiver-and-rainbow-inducing showers repeats itself seven or eight times. But not for a moment is there anything remotely *buidhe* about the lochan. I have known the place for forty years and never once have I looked in its earth-eye and thought, "Ah, the Yellow Lochan."

This particular Highland gem lies in a north-south mountain glen and in a wide tussock-grass arena the colour of... of... nope, nothing comes to mind at all. Its shade at the end of winter is a total absence of colour. I stared hard at it for several minutes, willing my eye to come up with something convincing, and when that failed utterly, just to come up with *something*. Finally I settled on snow. I decided that it looked as if it was dressed in a garment of snow that had been washed so often that the whiteness had gone out of it, so a kind of washed-out, bleached-grey snow, and even that suggests too specific a shade. Whether in bright sunlight or strafed by glowering sleet-shower clouds, nothing changed in the prevailing shade, because there was no shade to change. But it will be all the shades of spring green soon enough.

I have several reasons for returning to Lochan Buidhe. Firstly, it was right here and twenty years ago now that that old conversation with Don MacCaskill for a radio programme first introduced me to both the possibility of beavers returning to their Scottish heartlands, the possibility *and* the rightness of it. Throughout the research and the writing of this book I kept returning here in my head, because although at the time of that interview the great beaver adventure on which we are now embarked would not begin for several more years, it had provided me with a reference point when I went to Alaska and my wondering, wandering footfall imprinted a beaver landscape while I was looking for trumpeter swans.

And that also resonates with my long association with Lochan Buidhe. A narrow shoulder of land and a scrap of alder and willow wood are all that lie between the south shore of the lochan and a reed bed in a bay at the north end of Loch Lubnaig, and it is there that my love of swans found its voice in the shape of that mute swan nest site which I have watched for most of those forty years. And for much of that time, my writing was crowded with what I learned from the birds and their neighbours and fellow travellers and especially those birds-of-passage, the wild Icelandic whooper swans that grace every winter with their stopover visits on the loch. So long before Don and I and the Frenchman and a radio producer stood on this shore and talked about beavers, those reeds and the lochan and the narrow outlet burn that flows from one to the other were first the centre of my world, and then the centre of my writing world. So, from the moment the idea of this book was first proposed, the first thing that I had put in place was that Lochan Buidhe would be the end of the story, and a last homage to Don MacCaskill who had begun it for me. If I could also show in the most specific way that beavers had finally had the same idea as Don had dreamed, then I could write "*finis*" to signify the closing of a particularly satisfying circle, and I could elevate my memory of Don into the ranks of the immortals.

⊚⊚⊚

I shouldered aside the spindly, interlacing branches of two alders at the bottom of the oakwood slope and felt the land beneath my feet begin to slither and squelch and grow insubstantial, the blessed signature of part-time wetland. It gets more blessed and more insubstantial once you step into the wood that accommodates the narrow outlet burn from the lochan. The burn doesn't so much feed into the main loch but rather swithers among chaotic alders and willows, and sprawls into an untimely end in the reed bed where the swans nest. In the nesting season the cob bulldozes a channel the width of himself through the reeds from nest to burn, for although he spends most of his time out on the main loch, the lochan is part of his territory too, and he will defend it against all-comers, and his sudden flat-out emergence from burn into loch, head down and withdrawn, breast thrust forward like a bowsprit, wings arched like mainsails, the spectacle mirrored in the water to achieve twice the impact... all that is a sight to see, unless you are a stranger-swan at the far end of his black-eyed, unswerving gaze.

A drake goosander swam slowly into view, sitting preposterously atop his own perfectly realised upturned self, his sunlit reflection in the glassy shallows a yard from the lochan's shore. From snowy breast to the bottle-green gloss of his head to his red eye and scarlet bill, he was a stuff-strutter with a mirror in which to admire (not without some justification) the swanky hand nature had dealt him. In a particularly exotic embellishment of the moment, some irritation among the short feathers on the back of his head required the attention of a raised webbed foot, and when he lifted it out of the water it was the same outrageous shade of scarlet as the fish-devouring bill. He would become my constant companion over the next four hours in which he restlessly quartered every square yard of the watersheet and every yard of its shores. In sunlight he dazzled, in shadow and shower he simply glowed. I began to suspect him of reincarnation, camouflage for Don's ghost, checking out the validity of his dream.

I made for the burn. Just inside the edge of the wood, but with an open view over much of the lochan, there are stepping stones. Or at least there are rocks, and whether by accident or design, they permit a more or less dry-shod crossing, most of the time. *Most* of the time, because the other principal characteristic of the lochan and the reed bed is their proximity to the River Balvaig, and in winter and spring the Balvaig has a contemptuous relationship with its banks as snow-melt from a panorama of mountains and lesser hillsides feeds it far beyond what is reasonable. At the height of every winter I have known here, there comes a passage of days and sometimes weeks when lochan, reed bed, loch, river and usually the caravan site in the nearby village too, are a single sheet of water. One day, some old spring or other, a green plastic chair from someone's upstream patio drifted ashore precisely at the point where I was accustomed to sit (on the ground) and watch the nest across the reed bed. For three years thereafter I sat in comfort there until one more flood repossessed the chair and deposited it God knows where.

But now, on this sunlight-and-sleet day of late March, the water was comparatively low, the stones were dry, there was, as almost always, a fresh otter sprant on one of the rocks (almost always the same rock). I crossed, I eased off my small backpack, and made a more or less tolerable seat at the base of an alder trunk with an old mat, as I have done time after time for year after year. I sat and was still, and I let the day claim me as I had let a thousand other such days claim me.

But of course I was there on a different mission from a thousand other such days. I had come here, not just to the lochan but specifically to the outflow of the lochan, because I had reasoned that if beavers had returned, the first thing they would have done was dam the outflow. But there was not so much as a stick. All I knew about beaver presence in the area was that one had been seen on the River Balvaig upstream in Balquhidder, that one had been seen on the Leny near Callander, and that I had found that small dam and some gnawed bark, also near Callander. To get from the Balvaig to the Leny a beaver would have had to come this way, and to be aware at least that the lochan existed. Don and the Frenchman I also interviewed that day both thought the lochan would make perfect beaver habitat. Suddenly it looked like a flimsy hook on which to hang my hopes, even my expectations.

And yet, expecting to find a beaver presence here still seemed to make perfect sense. It's what they've been doing throughout their expansion across Tayside. I reminded myself of Louise Ramsay's observation after film of the Leny beaver appeared on YouTube in 2012: "Sightings like this show us how beavers would naturally move and disperse around the Scottish landscape, seeking out the very best habitats first before filling the gaps in between." And Lochan Buidhe is one of the best bits. The Frenchman thought so, a beaver expert who had been involved in reintroductions in France. Don had thought so, a nationally renowned forester and wildlife authority, to the extent that he had convinced national radio we should record him here. If nature is in the habit of entrusting its best interests to the dreams of human beings, it would surely be choosy about its choice of dreamers. More prosaically, Don's dream was surely the offspring of decades of working with forests and forest wildlife, a career in which he won a reputation for pushing the boundaries of the notably unmalleable Forestry Commission way of working, always craving a larger niche for nature within the Commission's essentially commercial operations. Sitting by the undammed burn (and for the first time, how like a beaver canal it suddenly looked to me!), I reassured myself of my complete faith in the dream.

I ate a late lunch sitting amid these so-familiar surroundings, amid so many, so familiar memories. One involved the only year the swans had responded to early nesting season floods by building a second nest on the west shore of the lochan instead of waiting for the floods to abate from the reed bed. Mute swans see their territory as a piece of water and they are very reluctant to fly during the nesting season, so the burn became a thoroughfare. One day I set up my camera to photograph them swimming through the flooded trees. They were completely accustomed to my presence by this time. It proved to be a long wait, but it is the kind of place where there is always something to watch. I wrote what happened next in *Waters of the Wild Swan*:

I had time to open eyes and ears to the swans' fellow travellers. The shrill giggling falsetto rising from the reed bed like smoke from a forest clearing was a familiar echo. There were little grebes here too. Tree pipits were in evidence for the first time that spring, free-falling cock-tailed down chutes of song, and the riverside trees behind me thrummed to the small orchestrations of tits and siskins. Green woodpeckers seemed to be calling from every compass point; eventually I pinned down four separate birds, but four green wood-peckers tend to sound like forty. A buzzard was displaying above a corner of a high hill wood, soaring and diving in pale imitation of eagles, but in the absence of eagles, impressive enough.

I dragged the glasses away from the buzzard, startled by a soft grunt a few yards away. The pen had sidled up the hidden shore of the lochan and was now swimming calmly past me through pools of vivid yellow light, threading a careful path between the trunks of half-submerged trees. She paused precisely where I had set up the camera in the hope that the birds might swim into a carefully composed frame of light and water and trees. I rattled through a handful of shots then stopped to admire the perfection of a wild swan in her wild element.

Trees gathered darkly about her, and the water gathered the same darkness beneath her, but the sun split the trees and shredded the darkness here and there with gold trimmings, a halo round a thin branch where it protruded from the water, a spark at the breast of the swan as she suddenly moved forward and pushed the water aside, a bright rim to the top of her head still wet from feeding. As she moved through the trees, their shadows rippled across her back and neck, and when she paused again in a yard of open water, she took the sun full on her and I never saw such a blaze of white. The cob caught up with her there and put twice the blaze on the floodwaters. I tried to think of something in nature that was more beautiful and nothing came to mind...

As I re-ran a film of that twenty-five-year-old memory through my mind, I suddenly became aware of Don's presence. I don't mean that I had a spiritual visitation (my mind doesn't work that way and I tend not to believe people whose minds do), but something in the particular nature of the circumstances contrived a vivid awareness of his company, as if I had just been talking to him. I have, of course, thought about him often in the fifteen years since he died, but suddenly I could see him and hear what the sound of his voice was like with renewed clarity. Thinking about it now, I suppose that some kind of heightened mental awareness is not so surprising given that I had invoked him and harnessed his dream to a mission of my own. But at that moment it rather knocked me sideways. My immediate response was to rebuke myself. What did I think I was doing, usurping a dead friend's dream and hitching it to my cause? But almost at once, a calmer, more rational voice – my own voice – reasoned that Don would have been delighted to know his dream had become reality. And I persuaded myself that, given the opportunity, he would have urged me to look beyond the undammed burn at my feet. So I stopped daydreaming about swans and started looking for souvenirs of beavers on the move. And I realised at that moment that from where I was sitting I could see a footprint in the mud at the water's edge, and although I was about three yards away and it was less than distinct I was instantly convinced it was a beaver footprint. I took my camera over to it to nail down the evidence, and of course it was an otter print. I rebuked myself for the second time in as many minutes and delivered a short lecture to myself, the gist of which was that I was not being very professional. An ever-so-slightly unchancy moment of renewed awareness of my old friend's voice was not going to conjure a beaver track at my feet. So I went to work.

I decided that I would walk down the riverbank all the way to the loch, examining the riverbank trees and the riverbank mud, and on the way back I would explore the little alder and willow thickets, and then I would scour as much of the shore of the lochan as I could reach, and if there was nothing at all to be found, then I would just keep coming back here until the day (surely not too distant) when there was a dam across the outlet burn and a lodge over there by the water's edge just below the alders just below the birches just below the oakwood.

And then I found it.

A solitary alder stood just a few yards from the riverbank. One of its limbs had been broken, broken but not severed from the parent tree, so that it angled down at about forty-five degrees and its spread of lesser branches hit the ground. On at least half of them the bark had been stripped and gnawed clean, and I recognised at once the bite patterns with which I had become so familiar on the Earn. I had also become familiar with how the exposed wood beneath the bark darkened over time from white through various darkening shades of brown and ultimately to grey. These stripped branches were mostly light-to-mid-brown but with many white teeth marks still showing. This was not a souvenir of 2012. This had happened within the last few weeks. I took some photographs for reference and walked on down towards the loch. And after about a hundred yards there was more of the same. This time the limb had broken from a bankside tree and lay partly prone on the grass and partly in the air above the top of the bank. There were no secondary branches, just a straightish limb. Beaver teeth had stripped several yards of bark, leaving a richly patterned calling card that looked vaguely oriental.

And then, at the very end of the riverbank where it yields at last to the open loch there was a patch of mud with one very clear footprint and one very unclear one, and the clear print showed the fainter imprint of the small front foot where the hind foot had stood on it. And this was not the work of the last few weeks, this was the work of the last few hours.

"There you are, Don," I muttered. And then, "You realise no-one will believe this when I write it down, don't you?"

Epilogue

LET WILDLIFE
MANAGE WILDLIFE

WILD SCOTLAND is a better place today than it was when Don MacCaskill and I did our radio interview on the shore of Lochan Buidhe, and the idea of wild Eurasian beavers at large in the landscape in the mid-1990s was nothing more than one man's sweet dream. Scotland is a better place because beavers have begun to reshape it, to reinvigorate it, to reinvent it, to cast their spell wherever they ply their architect's trade.

Beavers make a difference. Unlike other reintroductions we have attempted – or are likely to attempt this side of the wolf – they initiate and then sustain a rolling regime of constant landscape evolution and the associated liberation of literally endless opportunities for other creatures to thrive in their shadow. The Aigas Field Centre achieved a 400 per cent increase in biodiversity in five years. Imagine the ecological riches that would flow from a Scotland-wide beaver revolution that will last forever. Yet such is the nature and the potential of the animal to which we have finally hitched our star.

There is no going back now. We have also begun to dismantle those abject ecological calamities of the Victorian era and the other ruthless eras of our uncomfortably too-recent history that wiped from the map all those mammals whose presence challenged humanity's perceived right to tame wildness, to reorder nature and put it in its place so that it became the exploitable playground of the few, to manage wildlife so that it posed no challenge. As part of that debilitating process, the beaver was "managed" out of existence.

Let wildlife manage wildlife. It is the simplest of ideas. It is nature's way. It is also the genie in the bottle. And despite their desperate efforts to keep the cork jammed in the neck, the parties to the official trial in Knapdale, Argyll, have been forced – by nature itself – to re-evaluate that laborious, timid and top-heavy exercise in wildlife bureaucracy, in the face of the infinitely more fruitful, free-range, and bureaucrat-free re-introduction on Tayside. The original handful of escapees, possibly complemented by other unsanctioned releases, has established a widespread, vigorous and *wild* beaver population. It is not managed by the people who, one way or another, facilitated the planting of its seedbed in what has proved to be particularly fertile soil. It has been managed by the beavers.

Let wildlife manage wildlife. Beavers take wildlife management and landscape management out of our hands. Nature's architect is also nature's landscape architect, and for that matter, nature's usurper. Beavers do not adapt their way of life to fit the circumstances, they adapt the circumstances to fit the beavers' way of life. Specifically, they insist on wetland. Where it already exists, they safeguard it and expand it. Where it does not exist but they think it should, they move heaven and earth to make it happen. Where it has been drained and dried out by our own landscape-manhandling ways, beavers recognise the drained landscape for the fraud that it is, and they set about reviving and liberating the comatose marsh within.

Not everyone admires that approach. The most disgruntled are the ones who also manipulate the landscape for a living. You could be forgiven for thinking that, superficially at least, they should have much in common with beavers. But they are the drainers of fields, the realigners of rivers and streams, builders of flood defences against the realigned rivers and streams. In practice they are the opposite of beavers. Beavers are nature and what they do is nature's bidding. They manipulate landscape so that it maximises opportunity for as many creatures as possible, the exact opposite of the philosophy that motivates the field-draining, river realigning, flood-defence-building, opportunity-minimising tribe.

A mile or two to the north of Lochan Buidhe, the River Balvaig curls and uncurls dark and slow and deep through its natural flood plain, which is at its widest at the entrance into the beautiful Balquhidder Glen. For about ten years my writing base was within a few hundred yards of the place and it became a happy hunting ground for my nature-writing instincts. Autumn and winter storms would transform it, the Balvaig would fill to overflowing, and when it overflowed it created what looked for all the world like a loch, a square mile of water in which the course of the river was only identifiable by its riverbank trees. The locals call the phenomenon Loch Occasional.

The land is not used for anything else. The occasional misguided sheep finds its way there from nearby farms, but it is nature's place.

As soon as Loch Occasional begins to form, it becomes a wildfowl reserve. Ducks pour in – mallard, tufted, pochard, wigeon, teal, an occasional shoveller, twice that I know of a pintail. A small rabble of resident Canada geese goes wild with delight. But the show-stealers are the whooper swans – restless, itinerant, autumn-and-winter travellers from Iceland. From the moment they arrive in Scotland, usually in October or November, they drift from one regular haunt to another in family groups or in dozens, very occasionally in one or two hundreds. They are regular visitors to Loch Lubnaig and Lochan Buidhe to the south, Loch Earn just to the north, Loch Dochart over Balquhidder's northern mountains, and to Balquhidder's own lochs, Loch Voil and Loch Doinne. But almost as soon as Loch Occasional establishes its fleeting presence a flock comes winging round a mountainside and the air fills with Ellingtonian brasses, and my heart is inclined to turn over at the sound.

In spring, the land that was Loch Occasional off and on throughout the winter, is journey's end for migrating snipe, skylarks, oystercatchers, curlews, lapwings, tree pipits, grass-hopper warblers, reed buntings. The water table remains high and floods to a lesser extent at the least excuse to the delight of frogs, toads, dragonflies, damselflies. On the river, there are otters, kingfishers, dippers, grey wagtails, sandpipers, speculating ospreys.

Roe deer know the ways far out over the marshy ground, and red deer file down from hillside forests in the evening to browse the margins, and migrate back uphill towards the trees at sunrise, often lingering for a warm hour in the high fields.

If I were in charge of the Loch Lomond and the Trossachs National Park (a fate as unlikely as becoming an architect or an astronaut), I would be reintroducing beavers right there, because an inevitable consequence would be that Loch Occasional becomes Loch Semipermanent, and a glorious square mile of water and marsh would become something of a wildlife miracle. Add wolves into the mix and we could all start to take our national park system seriously.

There are other landscapes all across Scotland fit for beavers (remember John Lister-Kaye and Paul Ramsay identified 111 sites or prime beaver habitat north of the Tay). The most glaring opportunity is surely the Insh Marshes in the valley of the River Spey and in the shadow of the Cairngorms. They are already owned in their entirety by the Royal Society for the Protection of Birds, whose sphere of operations now extends far beyond birds to embrace all nature and all native habitats. The beaver is the most willing, the most accomplished, the most hospitable, and the most tireless of allies that nature conservation can summon to its cause, an architect that designs, re-designs, restores, and recreates wildness. For nothing. Forever.

And the beaver is capable of going beyond conservation. Aldo Leopold wrote in *A Sand County Almanac*:

> *The practices we now call conservation are, to a large extent, local alleviations of biotic pain. They are necessary but they must not be confused with cures.*

The beaver is capable of effecting cures, a capacity it shares with its brother wolf. The beaver's reintroduction is the precedent wild Scotland has been waiting for. The rein-troduction of the wolf is now surely no more than a question of time. The cork is out of the bottle. The genie has escaped into the eager embrace of the land. Let wildlife manage wildlife.

ACKNOWLEDGEMENTS

The seed from which this book eventually blossomed twenty years later was sown by the late Don MacCaskill. He was one of the wisest men I met in the matter of our relationship with nature, and he gave freely of that wisdom. He was also a source of great encouragement in my early years as a writer. He was an exceptional and visionary forester, an accomplished and award-winning nature photographer, and simply one of the best naturalists I have ever met.

There is also a profound debt of gratitude to be acknowledged to John and Lucy Lister-Kaye and to Paul and Louise Ramsay for generously sharing their expertise and the hospitality of their homes at Aigas and at Bamff, and for their patience in the face of my bombardment of questions. John and Paul are the true pioneers in the great adventure that has culminated in the reintroduction of beavers into Scotland; their own demonstration projects have been groundbreaking, and it is fair to say that without them, the professional nature conservation establishment would probably never have summoned the energy to conduct an official trial reintroduction. A word of thanks too to Alicia Leow-Dyke, former staff naturalist at Aigas, for answering still more of my questions.

I am particularly grateful to the staff at Mount Stuart, the ancestral home of the Marquess of Bute, for access to documents in the Mount Stuart archives, and for permission to quote from them. My visit there was invaluable in understanding the 3rd Marquess's beaver project on the island of Bute in the late 19th century.

To gain some kind of perspective of living with a cultural memory of beavers, I turned to a transatlantic friend, the nature writer and wildlife artist David M. Carroll from New Hampshire. This modest man's rare gifts were recognised with an American Genius Award by the MacArthur Foundation. Annie Dillard called him a national treasure. When I approached him about quoting some of his work he could not have been more enthusiastic and accommodating. The fact that I know David at all is because we were introduced by our mutual friend, the New Hampshire landscape painter Sherry Palmer, who has also been a constant source of news and information about beavers in America.

Nearer home, Mike Holliday and Polly Pullar have both contributed information and support from their wide range of practical nature conservation experience, and one way or another both have served the cause of this book well.

The experience or working with Saraband, the publishers of this book, is still a relatively new one for both writer and publisher, but it is already a relationship that I cherish. In what feels like no time at all, this will be our fourth book together. It is preceded by *The Eagle's Way* and the first two volumes – *Fox* and *Barn Owl* – of a series of small monographs entitled *Encounters in the Wild*, all of them published in the course of a very memorable 2014. So to Sara Hunt and Craig Hillsley and all at Saraband, thanks so much and it's good to be on board. And, as with *The Eagle's Way*, this book is graced by the cover artwork of Joanna Lisowiec, and the photographs of Laurie Campbell. This is the kind of company I like my writing to keep!

Finally, a particularly warm thank you to my agent Jenny Brown, who not only found a home for my work at Saraband, but whose efforts on my behalf are the main reason why my writing life is busier than ever.

ABOUT THE AUTHOR

JIM CRUMLEY has written more than thirty books, many of them on the wildlife and wild landscape of his native Scotland. He is a widely published journalist with regular columns in *The Courier* and *The Scots Magazine*, a poet, and occasional broadcaster on both radio and television.